Signal Processing

Signal Processing

Principles and Applications

D. Brook and R. J. Wynne

Edward Arnold

A division of Hodder & Stoughton

LONDON   BALTIMORE   MELBOURNE   AUCKLAND

094499

© 1988 D. Brook and R. J. Wynne

First published in Great Britain 1988

British Library Cataloguing Publication Data

Brook, D.
  Signal processing: principles and
  applications.
1. Signal processing
I. Title   II. Wynne, R. J.
  621.38′043   TK5102.5

ISBN 0–7131–3564–6

Typeset in 10/12 pt Times by Macmillan India Ltd, Bangalore 25.
Printed and bound in Great Britain for Edward Arnold, the
educational, academic and medical publishing division of Hodder
and Stoughton Limited, 41 Bedford Square, London WC1B 3DQ by
J. W. Arrowsmith Ltd, Bristol.

Preface

There are many books available on random signals and signal processing; all are good for their particular readership, but there are few directed to the undergraduate or the practising engineer whose formation occurred a generation or so ago, or to the technician engineer who has to learn about the subject. The advent of 'chip' technology has, of course, taken the subject of signal processing forward with it at a rapid rate and this book is intended for those who are early in their formation and those who are more mature, seeking a primer in the subject. It is hoped that it will also be useful for those working in non-engineering fields, such as medicine and biology, who need to know something of the subject.

The mathematics used is basic and is that commonly used in undergraduate engineering courses, so it is assumed that the reader is familiar with the use of the Laplace Transform, although an appendix briefly reviews the principles. The other transform methods used are the Fourier and z-transforms but the essential principles for an understanding of these is presented.

Emphasis is placed on understading the concepts rather than leading the reader through mathematical wizardry. In fact, in places, rigour is sacrificed to ease understanding. Concepts are developed using both discrete and continuous signals on an equal basis rather than regarding one as a special case of the other.

Certain examples are worked in considerable detail so as to give meaning to abstract ideas. It is so easy to assume that the reader is familiar with clever manipulations and to avoid explaining difficulties. Hopefully these worked examples will help the reader over the pitfalls.

This book has evolved over a number of years as a result of courses given at both Huddersfield Polytechnic and Manchester University. We are grateful to all undergraduates and those attending short courses who have contributed to the development of the book.

We wish to acknowledge the useful discussions and advice from colleagues on some of the applications, in particular T. A. Henry, B. C. Kwok, G. R. Tomlinson and J. T. Turner. Thanks are also due to Professor R. G. Barwick the Head of the Department of Engineering Systems, Huddersfield Polytechnic for the use of equipment to obtain experimental results.

Contents

1

Introduction

1.1 The Nature of a Signal

What is a signal? If we look in the Shorter Oxford Dictionary we find, amongst other meanings of the word, that signal means 'A sign or notice, perceptible by sight or hearing, given especially for the purpose of conveying warning, direction or information'. Although our use of the word goes beyond this rather restricted meaning, the essential ideas are present in this definition. 'A sign or notice, perceptible by sight or hearing' implies the presence of some kind of symbol, visible or audible whilst ' . . . conveying warning, direction or information' implies some form of message. The signals with which we will be concerned will include quantities which may or may not be visible or audible but which will certainly be detectable. Our signals will also convey some kind of message. Whilst the dictionary definition also implies an agreed code between sender and receiver we shall not be concerned with this aspect of the signal. Coding and the information content of signals is the province of information theory and is not within the scope of this book. We shall be concerned with the characterisation of signals and with the effect of systems on signals. We shall not be concerned with the origin of signals or their generation, indeed the origin of many signals is nature herself.

Examples of signals from nature are seismic waves, traffic noise and electrical noise. Typical communication signals are telegraph, telephone and television signals and signals generated for special purposes, such as test signals for electronic equipment or control systems.

So far we have not mentioned anything about the form that a signal could take. The dictionary definition implies that it is either a light wave or a sound wave, that is, it is either an electromagnetic wave or a wave of air pressure changes. Clearly signals can be other than these and for our purposes any quantity which varies over a period of time can be regarded as a signal. For example, voltage variations in an electrical circuit, the pressure variations in a hydraulic control system, the flow variations in the product of a process plant and the population variations in an insect colony are all signals.

The variation in the quantity regarded as the signal must be capable of being observed. Light and sound are directly observable by humans and animals using appropriate sensors or transducers, namely eyes and ears. Other quantities cannot be observed directly by humans (or animals) and thus require some intermediate sensor which converts the variation of the quantity into a form which is observable using available sensors. For example, to observe the pressure variations in a pipe containing fluid we must have equipment which will convert the pressure variations into, say, the movement of a spot on a cathode ray tube screen. Clearly the original signal is converted into another signal in a different medium. Every signal, of course, need not be converted into a form capable of direct observation by humans, however, conversion is often necessary and obviously depends on the apparatus observing the signal. It must be remembered that the transducer performing the conversion process must do so faithfully otherwise wrong conclusions can be drawn. An example of apparatus which needs to 'observe' a signal is a computer used in, say, process control where it may need to observe signals representing flow and temperature variations.

For our purposes we shall be concerned with the signal as a concept and only by way of example shall we regard it as the variation of a particular quantity. Incidentally, in engineering practice, signals very often appear as voltage variations because many sensors or transducers conveniently convert a quantity variation into a voltage variation. Also much control equipment and transmission equipment is electrically actuated.

We shall be concerned with describing signals formally, seeing how they are modified by systems through which they are transmitted and how to extract information from them about their source.

1.2   Signal Classification

1.2.1   Continuous Signals

If we are lucky enough to possess a barograph and have the self discipline to change the chart regularly then we can produce a continuous record of atmospheric pressure for an indefinite period. This record is that of a continuous signal, that is, it exists at every instant of time. Signals which have this property are known as continuous signals and in the jargon of the trade are often referred to as analogue signals. If their values in appropriate units (bars for atmospheric pressure) at some time t is represented by $x(t)$ then our record is a graph in the mathematical sense, Fig. 1.1.

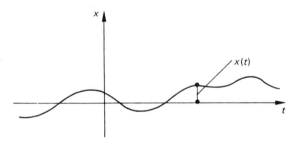

Fig. 1.1

Note that x is the symbol to represent the signal but that $x(t)$ means the *value* of x at time t.

1.2.2   Discrete Signals

If, however, we can only afford a barometer then, in order to observe atmospheric pressure over a long period, we must note its reading at intervals, say every hour. In this case our record would be a list of values and the times when the values were observed. This observation is an example of a sampled signal (we could regard the observation process as producing distortion or error). If we plotted these data they could be represented in either of the ways shown in Fig. 1.2.

An alternative name for this type of signal is digital signal but we shall use the term discrete signal. We have implied in Fig. 1.2 that the interval between samples is equal but

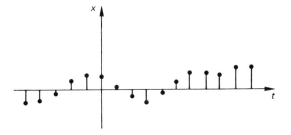

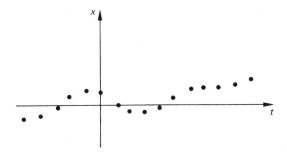

Fig. 1.2

this need not be so, however, in our studies we shall only concern ourselves with equal interval sampling, which is the most common anyway.

In many systems, instrumentation systems being representative ones, a continuous signal is the form of the input signal but it is necessary to sample it for manipulation purposes by, say, a computer. This produces discrete signals. It takes time however to effect the sampling process and the result is a train of pulses of finite time duration as shown in Fig. 1.3. If the pulses are very short in duration each varies little in height. If they are long they will vary in height and some satisfactory measure of height must be devised. Some typical measures are

1   the height at the beginning
2   the height at half duration or
3   the average height.

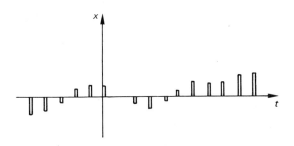

Fig. 1.3

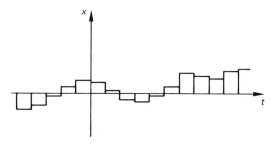

Fig. 1.4

Later we shall find it useful to use this kind of signal, especially in the case when the duration of the pulse is equal to the interval between samples. In this representation we will take the height of the pulse as being that of the signal at its beginning, Fig. 1.4.

1.2.3   Quantisation

Let us return to our barograph record and read what the pressure was at mid-day yesterday. We would find the line drawn by the pen to be about 0.5 mm wide and the smallest graticule on the scale to be 1 mm. Clearly, reason tells us that it would be pointless trying to read the chart to an accuracy better than 0.5 mm. If the total range of the chart was 100 mm then the most reasonable number of values readable across the range would be 201. Thus we have 'quantised' the barometric range represented by 100 mm into 201 possible values and this reading inaccuracy has effectively converted the original smooth signal into a stepped, or more correctly a quantised, signal, see Fig. 1.5.

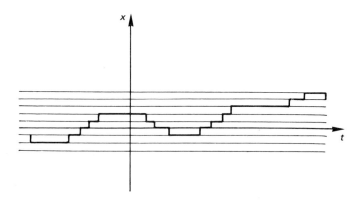

Fig. 1.5

The example cited is one in which the quantisation has occurred naturally as a result of the limitations of the apparatus and the means of observation; however, quantisation is often done deliberately for signal processing purposes.

Remember that quantisation arises when we perform calculations on values, either by hand or by computer, as a result of the rounding of numbers which obviously affects the accuracy of conclusion. We have implied equal quantisation intervals but, of course, this need not be so.

1.2.4   Discrete Signals and Quantisation

If we were to use our barograph with a computer for computational purposes then we would require a transducer which would convert the continuous signal of pressure variation into a quantised discrete signal. That is, we would have to sample the signal at intervals (normally equally spaced) and convert each value into a number. There would be a limited number of digits available for the representation because of the limited accuracy of the apparatus, hence the value would have to be quantised. The apparatus which performs this operation is an analogue to digital (A–D) converter or ADC, that is, it converts an analogue (continuous) signal into a digital (discrete–quantised) signal.

Fig. 1.6 shows such a signal. (In practice, for our example, there would be a transducer which would convert barometric pressure variations into variations of voltage and then this voltage would be sampled and quantised.)

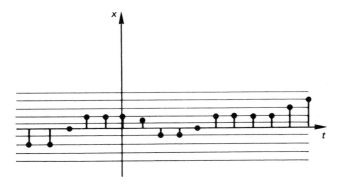

Fig. 1.6

1.2.5   Comments

It is obvious, but worth recording, that a signal which originates in continuous form and which is subjected to the above processes could suffer a loss in accuracy (or information content), thus it is important that sampling and quantisation should be done properly, otherwise wrong conclusions about the original signal may be drawn. It is worth noting at this stage that, in order to avoid loss of information due to sampling, the sampling rate must be at least twice that of the highest frequency present in the signal. This is known as the Nyquist sampling rate after Nyquist who proved this in the 1930's.

Further, the number of quantisation levels allowable depends on the accuracy required. We shall be concerned mainly with continuous signals, discrete signals and signals made up from pulses. (Figs. 1.1, 1.2, 1.3 and 1.4.)

1.2.6   Deterministic Signals

We all know that the electrical mains supply is alternating, that is, the voltage alternates between positive and negative values in a sinusoidal manner. It is possible to represent this sinusoidal variation by an equation of the following form:

$$v(t) = V_m \sin(2\pi f t + \phi).$$

This means that the voltage at the instant t is exactly determined and can be obtained from the equation. V_m is the greatest value reached and is a measure of the 'size' of the voltage, f is the frequency, that is, the number of cycles of the sine wave per second in Hertz (Hz). ϕ is a phase shift (radians) and its value depends on the relative position of the wave and the time chosen as origin. See Fig. 1.7.

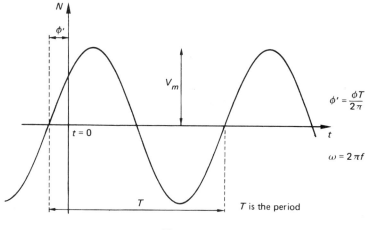

Fig. 1.7

Our equation clearly describes the voltage variation completely; it is possible to work out what its value will be at some time in the future as well as what it was at some time in the past.

Signals which can be described in this precise way are known as deterministic signals. Other examples are shown in Fig. 1.8.

Deterministic signals can be divided into three sub-groups: Periodic, Almost Periodic and Aperiodic. Periodic signals are ones which consist of a basic 'shape' of duration (period) T, which is repeated indefinitely. Fourier Series analysis tells us that such signals can be decomposed into a fundamental sine wave and a set (finite or infinite) of harmonically related sine waves. Almost periodic signals are ones composed of sine waves but which include components not harmonically related, that is, they are not related by rational numbers, for example:

$$x(t) = X_1 \sin(2\pi f_1 t + \phi_1) + X_2 \sin(\sqrt{2}\, 2\pi f_1 t + \phi_2) + X_3 \sin(\sqrt{3}\, 2\pi f_1 t + \phi_3).$$

Aperiodic signals are ones which do not have a repetitive form and are usually transient in nature, thus they are sometimes called transient signals. However, we will class Almost Periodic signals as Aperiodic.

1.2.7   Non-Deterministic (Random) Signals

Many signals which occur in nature, or which are generated for test purposes with the intent of simulating these natural signals, cannot be described by explicit mathematical equations.

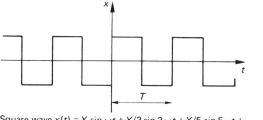

Square wave $x(t) = X \sin \omega t + X/3 \sin 3\omega t + X/5 \sin 5\omega t + \ldots\ldots$

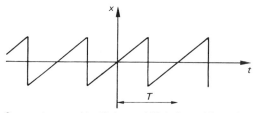

Saw tooth wave $x(t) = X \sin \omega t - X/2 \sin 2\omega t + X/3 \sin 3\omega t$

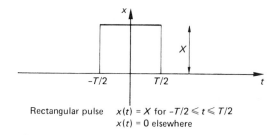

Rectangular pulse $x(t) = X$ for $-T/2 \leqslant t \leqslant T/2$
$x(t) = 0$ elsewhere

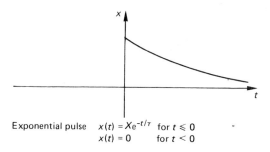

Exponential pulse $x(t) = X e^{-t/\tau}$ for $t \leqslant 0$
$x(t) = 0$ for $t < 0$

Fig. 1.8

These are known as non-deterministic or random signals. (Another term used is 'stochastic signals'). The description of random signals requires a different approach and will be the principal area of study of the next three chapters.

Examples of random signals are turbulence forces on airframes, ocean waves and the background noise of an audio system. Clearly in airframe design it is necessary to understand the characteristics and effects of the random turbulence induced forces; in the study of sea defences and erosion a study of wave characteristics is necessary and in the study of communication systems a knowledge of the extraneous noise is essential. These are only three examples but others come easily to mind. Typical records of random signals are shown in Figs. 1.9 and 1.10.

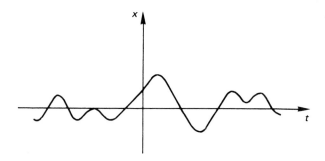

Fig. 1.9

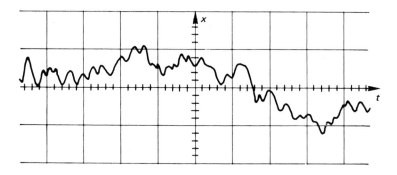

Fig. 1.10

We can classify random signals into two main groups, one of which can be further subdivided into two. We shall find that we need to describe these signals by statistical and probabilistic methods and those whose characteristics do not change with time are said to be stationary, while those whose characteristics do change with time are said to be non-stationary. Stationary signals are divided into ergodic and non-ergodic but we shall leave the meaning of these terms until later.

1.2.8   Summary of Signal Classification

Fig. 1.11 summarises the classification of signals

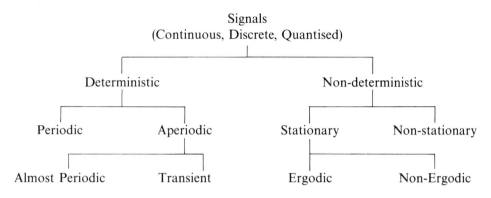

Fig. 1.11   Signal Classification

1.3   Averages

We begin our formal description of signals by examining certain basic characteristics which should be familiar to most readers but are included here for completeness and to set the scene for later work.

1.3.1   Mean Value

We all know that if we want to find the mean value (loosely known as the average value) of a set of values gathered in some measurement operation for example, we add them together and divide the sum by the number of values used. Thus the mean value of $N+1$ values of x is given by

$$\tilde{x} \simeq \frac{x_0 + x_1 + x_2 + \ldots + x_N}{N+1}$$

$$= \frac{1}{N+1} \sum_{i=0}^{N} x_i. \qquad (1.1)$$

This idea is applicable to signals in that the values of x could be observed values of a signal, (Fig. 1.12). Note that since a signal can take on both positive and negative values it is possible to have a mean value of zero, a sine wave is such a signal. Geometrically, the mean value can be regarded as the height of a rectangular signal enclosing the same area as the signal under consideration but with the same base duration, see Fig. 1.12.

Taking 17 values of the signal of Fig. 1.12 over the range shown we get

$$4, 3, -1, 2, 3, -3, -7, -5, 2, 8, 7, 4, 5, 8, 4, 0, 0$$

and an average value of 2 units. Note that we took a continuous signal and in observing the 17 values we effectively sampled the signal or converted it into a discrete signal. Further, in rounding the values we effectively quantised it. Also, the original continuous signal extended beyond the base duration in both directions. Clearly, and this is important to obtain greater accuracy for the mean value, we must make the duration of the sampling

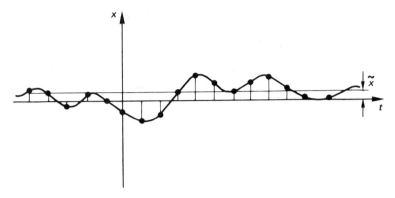

Fig. 1.12

interval short and make the base duration long, that is, N should be a large number and cover a large base duration.

To accommodate this mathematically we can modify equation (1.1) to

$$\tilde{x} = \underset{N \to \infty}{\text{Limit}} \frac{1}{N+1} \sum_{i=0}^{N} x_i. \tag{1.2}$$

We can regard this method of calculating the mean as a digital method and it is basically that used by digital equipment. However, we can define the mean using integration which is really only an infinite sum!

$$\bar{x} \simeq \frac{1}{T} \int_{t_1}^{t_2} x(t) . \, dt \tag{1.3}$$

where $\qquad\qquad T = t_2 - t_1$, see Fig. 1.13.

Clearly T, the base duration, should be as great as possible, so for mathematical precision, we define the mean value as

$$\bar{x} = \underset{T \to \infty}{\text{Limit}} \frac{1}{T} \int_{t_1}^{t_2} x(t) . \, dt.$$

For mathematical convenience we make $t_1 = -\dfrac{T}{2}$ and $t_2 = \dfrac{T}{2}$ giving

$$\bar{x} = \underset{T \to \infty}{\text{Limit}} \frac{1}{T} \int_{-\frac{T}{2}}^{\frac{T}{2}} x(t) \cdot dt. \tag{1.4}$$

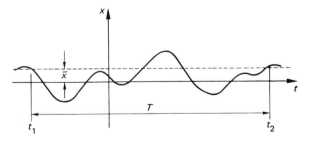

Fig. 1.13

To implement this expression we must know x as a function of time, thus it is only useful for calculation purposes with deterministic signals. However, the ideas embodied in the integral form of definition are extremely useful for developing concepts, as we shall see.

Example

A classic example is that of a sine wave—find the mean value of a sine wave, $x(t) = X \sin \omega t$.

We have
$$\bar{x} \simeq \frac{1}{T} \int_{t_1}^{t_2} X \sin \omega t \, dt.$$

For convenience let $t_1 = 0$ and then see the effect of using different values for t_2. With $t_1 = 0$ and $t_2 = T$ we get

$$\bar{x} \simeq \frac{X}{T} \int_0^T \sin \omega t \, . \, dt$$

$$= \frac{X}{T} \left[-\frac{\cos \omega t}{\omega} \right]_0^T .$$

(i) $T = \frac{1}{2}$ period $= \dfrac{\pi}{\omega}$

$$\bar{x} \simeq \frac{X \omega}{\pi} \left[\frac{(-\cos \pi) - (-\cos 0)}{\omega} \right]$$

$$= \frac{2X}{\pi}.$$

(ii) $T = 1$ period $= \dfrac{2\pi}{\omega}$

$$\bar{x} \simeq \frac{X \omega}{2\pi} \left[\frac{(-\cos 2\pi) - (-\cos 0)}{\omega} \right]$$

$$= 0.$$

Obviously an odd number of $\frac{1}{2}$ periods will result in $\bar{x} = \dfrac{2X}{\pi}$ and an integral number of whole periods will give $\bar{x} = 0$.

(iii) $T = \frac{9}{8}$ period $= \dfrac{9\pi}{4\omega}$

$$\bar{x} \simeq \frac{X \, 4\omega}{9\pi} \left| \frac{\left(-\cos \dfrac{9\pi}{4} \right) - (-\cos 0)}{\omega} \right|$$

$$= \frac{X \, 4\omega}{9\pi} \left[\frac{(-0.707) + 1}{\omega} \right]$$

$$= \frac{0.13 \, X}{\pi}.$$

If we use equation (1.4) we get

$$\bar{x} = \underset{T \to \infty}{\text{Limit}} \frac{1}{T} \int_{-\frac{T}{2}}^{\frac{T}{2}} X \sin \omega t \, dt$$

$$= \underset{T \to \infty}{\text{Limit}} \frac{X}{T} \left[\frac{-\cos \omega t}{\omega} \right]_{-\frac{T}{2}}^{\frac{T}{2}}$$

$$= \underset{T \to \infty}{\text{Limit}} \frac{X}{T} \left[\frac{\left(-\cos \frac{\omega T}{2} \right) - \left(-\cos \frac{\omega T}{2} \right)}{\omega} \right].$$

Whatever value T takes, $-1 \leqslant \cos \frac{\omega T}{2} \leqslant 1$ thus as T increases indefinitely $\bar{x} \to 0$. This trivial and well known example is cited to emphasize the point that the mean value depends on the base duration chosen.

Clearly it is important to handle data with care in order to get meaningful results. The wrong choice of data or the use of insufficient data can produce serious errors, a subject which will be taken up later.

1.3.2   Mean Square Value

The mean square value is the mean of the squares and the discrete data definition is

$$\widetilde{x^2} \simeq \frac{x_0^2 + x_1^2 + x_2^2 + \ldots + x_N^2}{N+1}$$

$$= \frac{1}{N+1} \sum_{i=0}^{N} x_i^2. \tag{1.5}$$

This is a biassed estimate but to get an unbiased one we should use $\dfrac{1}{N} \sum\limits_{i=0}^{N} x_i^2$. However, if N is large the error in using equation (1.5) is small. Most books on statistics would explain this.

The positive square root of this is the root mean square value (r.m.s value $= \sqrt{\widetilde{x^2}}$). In electrical science the power in a circuit is proportional to (voltage)^2 or to (current)^2, hence by analogy we say that the 'power' of a signal is (value)^2, or formally; for a discrete value of a sampled signal, instantaneous power $p = x_i^2$ and for an instantaneous value of a continuous signal

$$p = x^2(t).$$

Also the average power is

$$p = \widetilde{x^2}$$

or $\quad p = \overline{x^2} \quad$ (see the following example below).

It must be remembered that, in general, this is not true power. For example, if the signal is electrical current, then the true instantaneous power is $i^2(t)R$ where $i(t)$ is the instantaneous value of current at time t and R is the resistance through which i passes. In short when we use the term (value)^2 for signal power we are using the word power only as a convenience.

Example

Let us find the mean square value of the signal illustrated in Fig. 1.12.

Substituting the values into equation (1.5) gives $\widetilde{x^2} = 21.2$ and the r.m.s. value as 4.6 units. Again for accommodating mathematical precision we modify equation (1.5) to

$$\widetilde{x^2} = \underset{N \to \infty}{\text{Limit}} \frac{1}{N+1} \sum_{i=0}^{N} x_i^2. \tag{1.6}$$

For continuous signals we have

$$\overline{x^2} = \underset{T \to \infty}{\text{Limit}} \frac{1}{T} \int_{-\frac{T}{2}}^{\frac{T}{2}} x^2(t) dt. \tag{1.7}$$

Example

Find the mean square value of the saw tooth wave shown in Fig. 1.14.

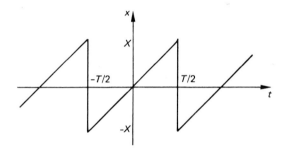

Fig. 1.14

We may use only one period as the base duration since the wave is periodic and an integral number of whole periods will give the same result (like the sine wave). Thus

$$\overline{x^2} = \frac{1}{T} \int_{-\frac{T}{2}}^{\frac{T}{2}} \left(\frac{2X}{T}\right)^2 t^2 dt$$

Since

$$x(t) = \frac{2X}{T} t \quad \text{for} \quad -\frac{T}{2} \leqslant x(t) \leqslant \frac{T}{2}$$

So

$$\overline{x^2} = \frac{4X^2}{T^3} \left[\frac{t^3}{3}\right]_{-\frac{T}{2}}^{\frac{T}{2}}$$

$$= \frac{4X^2}{3T^3} \left[\left(\frac{T^3}{8}\right) - \left(-\frac{T^3}{8}\right)\right]$$

$$= \frac{X^2}{3} \quad \text{(units)}^2.$$

The r.m.s value is $\dfrac{X}{\sqrt{3}}$ units.

1.3.3   Higher Order Averages

We can define 'higher order averages' as follows:

Mean cube value
$$\tilde{x^3} = \underset{N \to \infty}{\text{Limit}} \frac{1}{N+1} \sum_{i=0}^{N} x_i^3$$

or
$$\overline{x^3} = \underset{T \to \infty}{\text{Limit}} \frac{1}{T} \int_{-\frac{T}{2}}^{\frac{T}{2}} x^3(t) dt$$

Mean quartic value
$$\tilde{x^4} = \underset{N \to \infty}{\text{Limit}} \frac{1}{N+1} \sum_{i=0}^{N} x_i^4$$

$$\overline{x^4} = \underset{T \to \infty}{\text{Limit}} \frac{1}{T} \int_{-\frac{T}{2}}^{\frac{T}{2}} x^4(t) dt.$$

However, we shall not be concerned with these in this book, they are only mentioned for completeness.

1.3.4   Variance and Standard Deviation

Another important average is the variance. This is obtained by subtracting the mean value of the signal from each of the observed values and then finding the average of the squares of these differences.

Variance,
$$\sigma^2 = \frac{(x_0 - \tilde{x})^2 + (x_1 - \tilde{x})^2 + \ldots + (x_N - \tilde{x})^2}{N+1}$$

$$= \frac{1}{N+1} \sum_{i=0}^{N} (x_i - \tilde{x})^2. \tag{1.8}$$

The positive square root of this quantity, namely σ, is known as the standard deviation. Both variance and standard deviation are measures of the spread of the values around the mean value; the larger the spread the larger is σ^2 and hence σ.

For continuous signals this becomes

$$\sigma^2 = \underset{T \to \infty}{\text{Limit}} \frac{1}{T} \int_{-\frac{T}{2}}^{\frac{T}{2}} (x(t) - \bar{x})^2 dt.$$

It is obvious that there will be a relationship between mean square value and variance so we will develop it below:

$$\sigma^2 = \frac{(x_0 - \tilde{x})^2 + \ldots + (x_N - \tilde{x})^2}{N+1}$$

$$= \frac{(x_0^2 - 2x_0\tilde{x} + \tilde{x}^2) + \ldots + (x_N^2 - 2x_N\tilde{x} + \tilde{x}^2)}{N+1}$$

$$= \frac{x_0^2 + x_1^2 + \ldots + x_N^2}{N+1} - \frac{2\tilde{x}(x_0 + x_1 + \ldots + x_N)}{N+1} + \tilde{x}^2$$

$$= \tilde{x^2} - 2\tilde{x}\tilde{x} + \tilde{x}^2$$

$$= \tilde{x^2} - \tilde{x}^2.$$

This is the difference between the mean square value and the (mean value)^2.

1.4 More on Discrete Signals

We are used to the idea that, for continuous signals, $x(t)$ represents the value of the signal x at time t. Thus we need to develop a corresponding representation for discrete signals.

In the simplest case of such a signal there would be a collection of sampled values, such as

$$x(t_0), x(t_1), \ldots, x(t_i), \ldots \text{etc.}$$

or in a shorter form as

$$x[0], x[1], \ldots, x[i], \ldots \text{etc.}$$

$$\text{or} \quad x_0, x_1, \ldots, x_i, \ldots \text{etc.}$$

where we understand

$$x_i = x[i] = x(t_i)$$

which is an individual value (see Fig. 1.2).

Let us represent the collection, set, or sequence of values by $[x_i]$. For many calculations this set of values is not the most convenient form and we need to devise one suited to mathematical manipulation. It is important to realise that each of the above values occurs at a different time so it is not possible to form a signal representation by taking their sum, thus it is necessary to have a representation which preserves their time spacing and to do this we make use of the function, $\delta[i]$ (sometimes known as the Kronecker delta but we will call it the unit pulse function). This is defined as

$$\delta[i] = 1 \qquad \text{for} \quad i = 0 \tag{1.9}$$

$$\delta[i] = 0 \qquad \text{for} \quad i \neq 0$$

see Fig. 1.15.

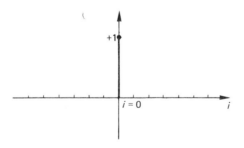

Fig. 1.15

Such a function which is shifted on the time axis by n sampling intervals (equal intervals) as illustrated in Fig. 1.16 would be written as

$$\delta[i - n]$$

which means

$$\delta[i - n] = 1 \qquad \text{for} \quad i = n$$

$$\delta[i - n] = 0 \qquad \text{for} \quad i \neq n. \tag{1.10}$$

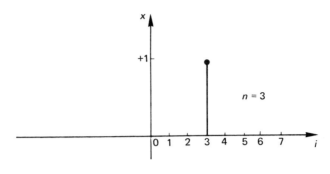

Fig. 1.16

Now, instead of representing the signal x by a set of values we can represent it by the sum

$$x^*[i] = x[0]\delta[i] + x[1]\delta[i-1] + \ldots + x[n]\delta[i-n] + \ldots$$

or

$$x^*[i] = \sum_{n=0}^{x} x[n]\delta[i-n]. \tag{1.11}$$

Likewise we need a representation for signals of the form shown in Fig. 1.3 and to do this we need the concept of the impulse function.

Recall that $\delta(t)$ is the symbol for a pulse (an ideal one!) which occurs at $t=0$ and has zero time duration, infinite height and unit area (or strength), see Fig. 1.17.

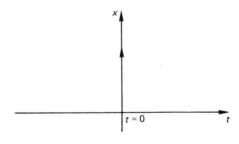

Fig. 1.17

Mathematically this is defined as follows:

$$\delta(t) = \infty \qquad \text{for} \quad t = 0$$
$$\delta(t) = 0 \qquad \text{for} \quad t \neq 0 \tag{1.12}$$
$$\int_{-\infty}^{\infty} \delta(t)\mathrm{d}t = 1.$$

To help us visualise this pulse, whose full name is unit impulse, we can consider it to be the limiting case of a pulse which has finite width, or time duration τ, and given height, h. Let the duration be halved to $\dfrac{\tau}{2}$ and the height doubled to $2h$ leaving the same area or

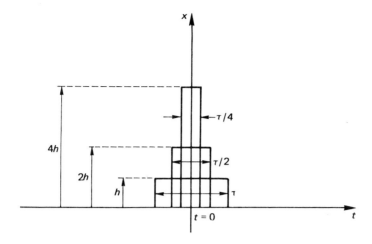

Fig. 1.18

strength, $h\tau$. Let the duration be halved again and the height doubled again leaving the strength the same. Continue this process indefinitely and we get a pulse of zero width and infinite height but strength $h\tau$. When $h\tau = 1$, it is dimensionless (meaning that h is dimensionally (Time)$^{-1}$) we have a unit impulse, Fig. 1.18. By definition $\delta(t)$ occurs at $t = 0$, thus an impulse occurring at $t = \eta$ would be represented by $\delta(t - \eta)$ and equation (1.12) becomes

$$\delta(t - \eta) = \infty \qquad \text{for} \quad t = \eta$$

$$\delta(t - \eta) = 0 \qquad \text{for} \quad t \neq \eta \tag{1.13}$$

$$\int_{-\infty}^{\infty} \delta(t - \eta)\,\mathrm{d}t = 1.$$

It is important to appreciate the distinction between the delta function $\delta[i - n]$ and the impulse function $\delta(t - \eta)$; the latter has a strength as measured by its area whereas the former does not.

The concept of strength allows us to use the impulse to represent a pulse of finite duration as follows. Take a single pulse of height $x(\eta)$ occurring at $t = \eta$ but of short duration $\Delta\eta$ so its strength will be $x(\eta) \cdot \Delta\eta$. This is then regarded as a multiplier for a unit impulse resulting in the representation

$$x(\eta) . \Delta\eta . \delta(t - \eta) \tag{1.14}$$

for the finite pulse.

A train of pulses can then be expressed as the sum

$$x^*(t) = x(0) . \Delta\eta . \delta(t) . + x(1)\Delta\eta\delta(t - 1) + \ldots + x(\eta) . \Delta\eta . \delta(t - \eta) \tag{1.15}$$

$$= \sum_{\eta = 0}^{\infty} x(\eta) . \delta(t - \eta) . \Delta\eta. \tag{1.16}$$

(Note that we use the asterisk to indicate that the signal is discrete and made up from pulses of finite duration and strength.)

This representation is also valid when the pulse duration is the same as the sampling interval provided it is small. It is also worth noting that, if this interval is infinitesimally small so that $\Delta\eta \to d\eta$ and the summation tends to an integral, $x^*(t) \to x(t)$

$$x^*(t) \to x(t) = \int_0^\infty x(\eta) . \delta(t-\eta) . d\eta. \tag{1.17}$$

A word about the limits is in order here. In both equations (1.11) and (1.16) we assumed that x began at $t = 0$ and continued to infinity but if it began at any other time or finished at a finite time then the limits would be altered accordingly. Commonly occurring cases are

$$x^*[i] = \sum_{n=-\infty}^{\infty} x[n] . \delta[i-n],$$

$$x^*[i] = \sum_{n=0}^{k} x[n]\delta[i-n] \tag{1.18}$$

with corresponding forms for equations (1.16) and (1.17). Remember that $x_i = x[i]$ is the value of the ith sample whereas $[x_i]$ is the set, collection or sequence of values.

Exercises 1

1   For the voltage wave shown in Fig. 1.19 find the mean value and the rms value. (Hint: either draw to scale, choose a small sampling interval and obtain a set of values, or find the net area and divide by the duration).

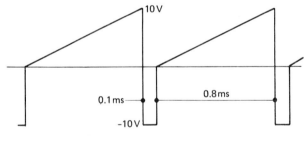

Fig. 1.19

2   The data below is taken from a random signal and the interval between each value is 1/20 sec. Find the mean value and the rms value.

−6	52	58	49	53	12	31	77	56	55	29	12	−40	−43	−38	−18	−32	−36	3	−29
−79	−83	−47	10	45	36	11	−18	−20	29	18	−39	−37	−69	−40	−28	−38	−82	−91	−65
−36	8	1	−1	23	67	46	−6	−46	−57	−77	−79	−27	−3	3	18	27	8	44	96
98	88	42	34	16	−43	−37	−15	−37	−58	−69	−62	−66	−86	−42	31	31	18	40	18
−16	9	35	30	61	77	23	−44	−41	24	37	32	73	51	22	2	31	52	67	35
48																			

3   Use equation (1.7) to show that the rms value of the sine wave, $x(t) = X \sin \omega t$, is
$$\sqrt{\overline{x^2}} = \frac{X}{\sqrt{2}}$$

4   Draw one period of $x(t) = 10 \sin \omega t$ to scale and read off the following number of samples for the period.

(a) 11 includes zeroth and last value
(b) 21 includes zeroth and last value
(c) 51 includes zeroth and last value
(d) 101 includes zeroth and last value

For each set of samples find $\tilde{x}$ and $\sqrt{\overline{\tilde{x}^2}}$ and compare with the true value $\tilde{x} = 10/\sqrt{2}$.

5   A signal z is made up from constituent signals w, x, and y in the following way: $z = w + x - y$ in which w, x, and y are each sinusoidal but of different frequency. Show that the mean square value of z is the sum of the mean square values of the constituent signals that is $\overline{z^2} = \overline{w^2} + \overline{x^2} + \overline{y^2}$.
(Hint: form the intergrad of equation (1.7) as $[W \sin \omega_1 t + X \sin \omega_2 t - Y \sin \omega_3 t]^2$ and complete the intergration).

2

Amplitude Characterisation

2.1 Introduction

In Section 1.2.6 we discussed the general ideas of deterministic signals and from this brief discussion it is easily seen that the explicit equation which describes such a signal gives us all the information relevant to it. It tells us about its amplitude, its position in relation to a time origin, its shape and about the way in which it changes with time. For example, in the case of a sine wave we have

$$x(t) = X\sin(\omega t + \phi)$$

X is a measure of the 'size'

$\omega = 2\pi f$ is a measure of the rate at which changes occur with time

$\sin(\omega t + \phi)$ tells us about the shape

ϕ tells us about the position in relation to the time origin.

We said that a random signal cannot be described in this way and we can only find out what its value was at some time in the past by examining a record. We have no means of finding out exactly what its value will be at some time in the future; however, we may be able to estimate its probable value if we know something about its statistics. We shall examine this problem now but we will need to know (or revise) something about elementary probability. To many this might be a closed subject or one which has given difficulties but it is hoped that the following sections will be clear enough to promote a basic practical understanding.

2.2 Basic Properties of Probability

If we throw a balanced die we expect, intuitively, that the chances of getting a 'one' (or any number from one to six inclusive) is one in six simply because there are six faces on the die and any one can come up with the same chance. Try this out: throw a die a large number of times (≈ 100) and record each result. Count how many times each of the faces came up.

Let us take another example. Suppose we are inspecting the output of a machine which is cutting a long steel bar into smaller lengths. It is required that the lengths of rod shall normally be 100 mm and that we have measured 50 such pieces. The results are listed below.

100.2, 99.1, 99.5, 98.1, 99.2, 99.8, 99.9, 100.5, 100.2, 100.1, 101.0, 100.8, 98.2, 98.5, 101.8, 101.9, 98.0, 100.5, 100.9, 98.6, 99.9, 100.1, 100.1, 100.2, 99.8, 99.7, 98.2, 98.3, 99.5, 98.7, 98.8, 100.4, 100.7, 101.5, 101.8, 99.1, 98.4, 99.5, 100.0, 99.0, 98.9, 99.1, 99.4, 99.9, 101.9, 100.5, 100.1, 100.0, 99.4, 99.2. Now classify these results into groups.

Table 2.1

Group	98.0–98.9	99.0–99.9	100.0–100.9	101.0–101.9
Number in Group	11	17	16	6

On the basis of this sample of 50 we could say with a probability of $11/50 = 0.22$ that if we selected a rod from the production run it would have a length between 98.0 and 98.9 mm inclusive. The results also show us that we are more likely to pick up a rod having a length between 99.0 and 99.9 than one from any other group because the probability for this group is $17/50 = 0.34$ whilst the other groups have probabilities of $11/50 = 0.22$, $16/50 = 0.32$, $6/50 = 0.12$.

Obviously the more rods we inspected, that is, the bigger the sample taken, the more accurate our probability estimate would be.

This idea is the reasoning behind the concept of 'frequency probability', that is, a probability definition based on frequency of occurrence or, in short, on number counting. Formalising this we get the following definition. Let the number of measurements falling in a specified group be n and let the total number of measurements in the test be N.

Then the probability of a measurement in this group is defined as:

$$P(\text{specified group}) \simeq \frac{n}{N}.$$

As already stated the greater N is, and hence n, the more reliable the probability measure. More accurately we can write

$$P(\text{specified group}) = \underset{N \to \infty}{\text{Limit}} \frac{n}{N}.$$

Two simple properties of probability are deducible from the definition:
(1) Probability is always positive. Obvious since both n and N are positive numbers.
(2) Probability is never greater than unity. Again obvious since n will never be greater than N.

These two stated mathematically are

$$0 \leqslant \text{Probability} \leqslant 1.$$

Refer again to our rod inspection example. We notice that the probability of picking a rod whose length lies between 98.0 and 101.9 mm is $50/50 = 1.0$, that is

$$P\,(98.0 \text{ to } 101.9) = \frac{11 + 17 + 16 + 6}{50} = 1.$$

Writing this differently we get

$$P(98.0 \text{ to } 101.9) = \frac{11}{50} + \frac{17}{50} + \frac{16}{50} + \frac{6}{50} = 1$$
$$= P(98.0 \text{ to } 98.9) + P(99.0 \text{ to } 99.9)$$
$$+ P(100.0 \text{ to } 100.9) + P(101.0 \text{ to } 101.9).$$

We can generalise this as follows. Consider some test (it could be the inspection problem) in which the total number of observations is N. Let there be n_1 observations in group 1, n_2 in group 2 and so on up to n_r in group r. Then

$$P \text{ (group 1)} = n_1/N$$
$$P \text{ (group 2)} = n_2/N$$

$$P \text{ (group } r) = n_r/N.$$

Obviously $n_1 + n_2 + \ldots + n_r = N$ because this is how we formed the grouping. Now

$$\text{Prob (group 1)} + \text{Prob (group 2)} + \ldots \text{Prob (group } r)$$

$$= \frac{n_1}{N} + \frac{n_2}{N} + \ldots + \frac{n_r}{N} = 1. \qquad (2.1)$$

As an exercise reclassify the data above into the following new groups.

98.0 to 98.4, 98.5 to 98.9, 99.0 to 99.4, 99.5 to 99.9
100.0 to 100.4, 100.5 to 100.9, 101.0 to 101.4, 101.5 to 101.9.

Find each group probability and check with equation (2.1). We can draw a chart to represent grouping. Return to the original groups. On the horizontal axis set out the group intervals (Fig. 2.1), draw a rectangle on each interval so that its area is proportional to the relevant probability. (In this case the height of each rectangle will be proportional to probability because the horizontal bases are equal.)

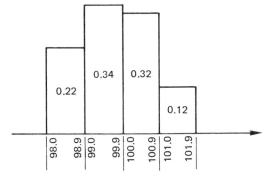

Fig. 2.1

This is known as a histogram and very readily tells us how the probabilities are distributed. We see that there is greater probability of picking a rod around 100 mm long than, say, around 101 mm or 98 mm. What is the total area of the histogram? Draw the histogram for the regrouped data and find its area.

Our example has used only 50 elements of data but it is obvious that if we had been dealing with much greater numbers, say 5000, we could have arranged to have much finer grouping and also to achieve better probability estimates and a much smoother histogram.

2.3 Probability and Signals (Discrete)

There is, of course, much more to the theory of probability than that described above but it should be adequate to appreciate basic ideas of probability relevant to signals. Consider a continuous signal which has been sampled as shown in Fig. 2.2 for which the discrete values obtained have been recorded.

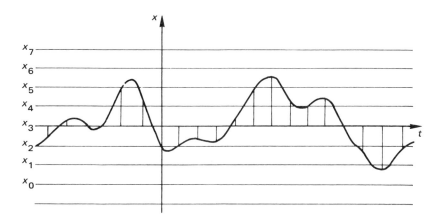

Fig. 2.2

Now classify into groups in the same manner as before. Let the total number of data values be N and let the number in each group be $n_1, n_2, \ldots n_r$. We could draw a histogram of the data and get a diagram of the distribution of the signal values; the probabilities would represent a measure of how often the signal could be expected in the respective intervals and the histogram would be a graphical representation of this. Indeed the data in Section 2.2 might be that obtained in this way for, say, the voltage output of a transducer and Fig. 2.1 would be its graphical representation.

We can, of course, regard the measurements of rod length as values of a discrete signal representing the output of the production process and this signal would look rather like Fig. 2.3.

Fig. 2.3

In short we are saying that probability distribution is a measure of the signal size and says something about its shape, better still it says something about where we are most likely to find a signal value if we take a random instantaneous measurement.

Having seen how the simple ideas of frequency probability can be applied to signals in a fairly loose way let us now look more closely at this application. Refer to Fig. 2.2. Not only have we sampled the signal but we have defined a number of quantisation levels. Again assume we have recorded the sample values (actual values, not quantised values). Now take quantisation level x_0 and count the number of signal values which are equal to or less than this level. Let the number be n_0. Take quantisation level x_1 and again count the number of signal values equal to or less than x_1. Repeat the counting for all quantisation levels up to the highest which will be chosen to exceed all expected values of the signal. For the signal of Fig. 2.2 we get $n_0 = 0$, $n_1 = 1$, $n_2 = 5$, $n_3 = 10$, $n_4 = 13$, $n_5 = 18$, $n_6 = 20$, $n_7 = 20$. The total number of values is 20. On the basis of this sample we can say:

$$P(x \leqslant x_0) = 0$$
$$P(x \leqslant x_1) = \tfrac{1}{20} = 0.05$$
$$P(x \leqslant x_2) = \tfrac{5}{20} = 0.25$$
$$P(x \leqslant x_3) = \tfrac{10}{20} = 0.5$$
$$P(x \leqslant x_4) = \tfrac{13}{20} = 0.65$$
$$P(x \leqslant x_5) = \tfrac{18}{20} = 0.9$$
$$P(x \leqslant x_6) = \tfrac{20}{20} = 1$$
$$P(x \leqslant x_7) = \tfrac{20}{20} = 1.$$

We can now plot a graph showing probability against quantisation level in the form of a bar chart, see Fig. 2.4. Since we can regard the data in Section 2.2 as signal data (and for the present purposes the nature of the signal is irrelevant) we can produce a cumulative probability distribution graph as above. Choose the quantisation levels as 97.9, 98.9, 99.9, 100.9, 101.9 and classify in these groups. (Table 2.2).

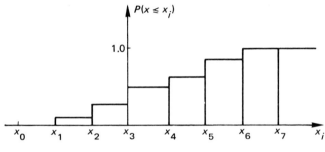

Fig. 2.4

Table 2.2

Group	No. in Group	Probability
97.9	0	0
98.9	11	0.22
99.9	18	0.56
100.9	44	0.88
101.9	50	1.0

Fig. 2.5 is a bar chart representing this data.

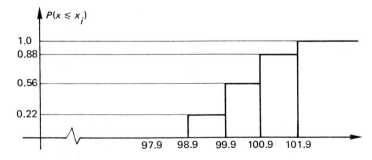

Fig. 2.5

This also represents the way in which the signal level is distributed but not, perhaps, quite so effectively as the histogram type of representation. The latter form of representation is called the probability density distribution for reasons which will be apparent shortly. Figs 2.6 and 2.7 show a Cumulative Probability Distribution (CPD) and a Probability Density Distribution (PDD) for a typical signal for which the quantisation level intervals are fairly small. Clearly if the intervals were infinitesimally small the graphs would be smooth curves.

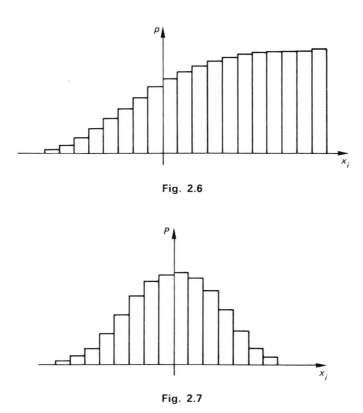

Fig. 2.6

Fig. 2.7

The ordinates of Fig. 2.6 have dimensions of probability and we have deduced that the total area under a PDD curve (histogram) is unity, but we have not yet discussed the dimensions of the ordinates of the latter graph. We know that the area of a vertical strip of the PDD graph is dimensionally probability, that is, the area of the strip between ordinates representing quantisation levels x_i and x_{i+1} is the probability that the signal level is less than or equal to x_{i+1} but greater than x_i. Since the abscissa is dimensionally that of the signal, (e.g. volts or $m^3 s^{-1}$ etc.), it follows that the ordinates will have dimensions of probability/ signal unit. The reason for the name probability density is now clear and we speak of the ordinates being probability density. We will abbreviate and use P for probability and p for probability density.

2.4 Properties of Probability Distributions

Refer to Fig. 2.8 and recall that the area of a strip in the PDD graph is the probability that the signal will be found greater than the lower boundary ordinate, x_i, and less than or equal

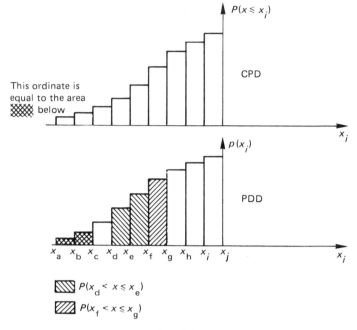

Fig. 2.8

to the upper, x_{i+1}. It follows that the area of two adjacent strips will be the probability of finding the signal greater than x_i but less than or equal to x_{i+2} and so on for n adjacent strips.

If we take all strips from the left (x_a) up to, say, x_h we deduce that the area of all these strips is $P(x_a < x \leqslant x_h)$. Since the probability that the signal is less than or equal to x_a is zero, that is, there is zero area to the left of x_a we can say that the area under the curve from $-\infty$ up to x_h is the probability that the signal is less than or equal to x_h. This probability is the height of the ordinate at x_h on the CPD graph. Clearly the two diagrams are simply related: the height of an ordinate on the cumulative graph is equal to the area under the distribution curve to the left of that ordinate.

2.5   Probability and Signals (Continuous)

In practice when performing observations on signals it is impossible to digitally process data using infinitesimally small intervals for sampling and quantisation; however, for developing certain principles it is valuable to regard signals as being sampled and quantised in this way. Clearly for sampling and quantisation in such a way the CPD and PDD curves would become smooth as shown in Fig. 2.9.

We can express the ideas of section 2.4 very concisely using calculus:

$$P(x \leqslant x') = \int_{-\infty}^{x'} p(x)\mathrm{d}x \tag{2.2}$$

$$p(x') = \frac{\mathrm{d}}{\mathrm{d}x} P(x \leqslant x') \tag{2.3}$$

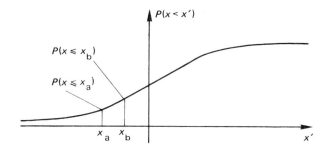

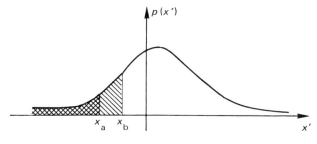

Fig. 2.9

x is any value of the signal and x' is a particular value and since the two curves are now smooth $P(x \leqslant x')$ and $p(x')$ can be regarded as functions. Equation (2.2) states simply that the area under the PDD curve from $-\infty$ to a particular value, x' is equal to the probability that the signal value will be less than or equal to x'. Equation (2.3) is a simple result of equation (2.2) using basic calculus principles.

We can also say

$$P(x_a < x \leqslant x_b) = \int_{x_a}^{x_b} p(x)\mathrm{d}x$$

$$= P(x \leqslant x_b) - P(x \leqslant x_a) \qquad (2.4)$$

in which x_a and x_b are two values of x'. This is easy to prove: .

$$P(x \leqslant x_a) = \int_{-\infty}^{x_a} p(x) \cdot \mathrm{d}x$$

$$P(x \leqslant x_b) = \int_{-\infty}^{x_b} p(x) \cdot \mathrm{d}x$$

and

$$P(x \leqslant x_b) - P(x \leqslant x_b) = \int_{-\infty}^{x_b} p(x)\mathrm{d}x - \int_{-\infty}^{x_a} p(x)\mathrm{d}x$$

$$= \int_{x_a}^{x_b} p(x).\mathrm{d}x$$

$$= P(x_a < x \leqslant x_b).$$

Note that $p(x')\mathrm{d}x \backsimeq p(x')\Delta x$, the area of the strip over Δx, is $P(x' - \Delta x) < x \leqslant x'$. Also note

$$\int_{-\infty}^{\infty} p(x)\mathrm{d}x = 1. \qquad (2.5)$$

2.6 Another way of Estimating $P(x \leqslant x')$ for Continuous Signals

We have seen how to estimate $P(x \leqslant x')$ by observing instantaneous values of a signal (discrete or continuous) and applying simple ideas of frequency probability. Another way of finding $P(x \leqslant x')$ is by measuring time intervals as follows. Take a continuous record of the signal under study, such as that shown in Fig. 2.10. Choose quantisation levels as before but now we measure, for each level, the time for which the signal is equal to or less than a given level. For example in Fig. 2.10, we see that for level x_i this time is $t_i = \tau_1 + \tau_2 + \tau_3 + \tau_4 + \tau_5$. Let T be the duration of the signal sample, then we can estimate $P(x \leqslant x_i)$ as

$$P(x \leqslant x_i) \simeq \frac{t_i}{T}.$$

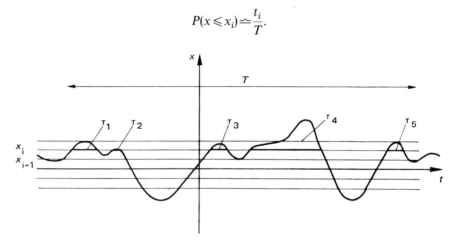

Fig. 2.10

This measurement can be repeated for all relevant quantisation levels and so we can obtain a large number of probability values, $P(x \leqslant x_{i-1})$, $P(x \leqslant x_i)$, $P(x \leqslant x_{i+1})$ etc. from which a CPD graph can be plotted. Very closely spaced levels would clearly give a large number of values to plot and thus result in a smooth CPD curve.

Ideally we should make $T \to \infty$, that is, our signal sample should be of infinite duration, in order to get accurate values for $P(x \leqslant x_i)$. As before, to accommodate this concept, we use

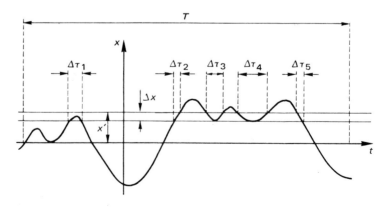

Fig. 2.11

the limit notation:

$$P(x \leqslant x_i) = \underset{T \to \infty}{\text{Limit}} \frac{t_i}{T}.$$

Recalling equation (2.3) $(p(x') = \frac{d}{dx} P(x \leqslant x'))$ we realise that the PDD graph is a plot of the slope of the CPD graph, hence, having obtained the latter by the above procedure we can derive the PDD graph for the signal. However, it is possible to estimate $p(x')$ directly by invoking the principle that

$$p(x')\Delta x = P(x' - \Delta x < x \leqslant x').$$

Refer to Fig. 2.11 where we see that, for the sample of duration T, the time for which the signal lies in the increment Δx at level x_i is

$$\Delta t_i = \Delta \tau_1 + \Delta \tau_2 + \Delta \tau_3 + \Delta \tau_4 + \Delta \tau_5.$$

We can now say $P((x' - \Delta x) < x \leqslant x') \simeq \dfrac{\Delta t_i}{T}.$

If we repeat this measurement for increments Δx at different levels we can obtain a set of values for $p(x')\Delta x$ for a relevant range of x'. If Δx is the same at each level we can then plot a PDD graph for which the ordinate at x' is proportional to

$$p(x') \cdot \Delta x = P((x' - \Delta x) < x \leqslant x') \simeq \frac{\Delta t_1}{T}.$$

Table 2.3

Name	$p(x)$		Curve
Normal or Gaussian	$\dfrac{1}{\sqrt{2\pi}\sigma} \exp\left[-\dfrac{1}{2}\left(\dfrac{x-\mu}{\sigma}\right)^2 \right]$		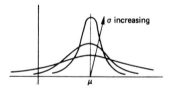
Rayleigh	$\dfrac{x}{\sigma^2} \exp\left[\dfrac{-x^2}{2\sigma^2}\right]$ $p(x) = 0$	for $x \geqslant 0$ for $x < 0$	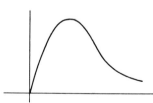
Maxwell	$\dfrac{x^2}{\sigma^2}\sqrt{\dfrac{2}{\pi}} \exp\left[\dfrac{-x^2}{2\sigma^2}\right]$ $p(x) = 0$	for $x \geqslant 0$ for $x < 0$	

μ is the mean value and σ^2 is the variance

2.7   Mathematical Probability Distributions

An intriguing aspect of this description is that for many naturally occurring random signals which, on looking at a record (e.g. Fig. 1.10), seem to be quite disordered, yield probability distributions which are relatively simple mathematical curves. Table 2.3 indicates certain common types but the reader who is interested in their detailed derivation and/or origin is referred to books on statistics and probability and on advanced books on random signals. We will concern ourselves only with the engineering significance.

2.8   Probability Distributions for Deterministic Signals

Whilst the illustrations and examples above have used random signals there is no reason why the CPD's and PDD's cannot be obtained for deterministic signals. However, since deterministic signals can be described in other ways probability distributions are less important for these cases. Nevertheless, it is useful to derive these distributions for typical signals to gain insight into the concepts.

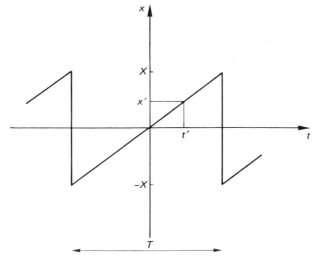

Fig. 2.12

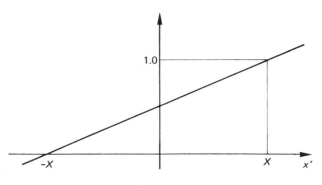

Fig. 2.13

Example

The CPD and PDD for a saw-tooth (or triangular) wave. See Fig. 2.12. Since the signal is periodic we need only consider one period in the solution. For this one period we have

$$x = \frac{2X}{T}t.$$

Choose an arbitrary level x' for which we can say, using the ideas of section 2.6.,

$$P(x \leqslant x') = \frac{\dfrac{T}{2} + t'}{T}$$

$$= \frac{1 + \dfrac{x'}{X}}{2}.$$

The graph of this function is shown in Fig. 2.13. However, we know that x can never be less than $-X$ or greater than X; therefore $P(x \leqslant -X) = 0$ and $P(x \leqslant X) = 1$. Clearly the graph of $P(x < x')$ will be that of Fig. 2.13 truncated as shown in Fig. 2.14(a). Note that we now use x' to conform with our usual notation. If we invoke equation (2.3), $p(x') = \dfrac{d}{dx} P(x \leqslant x')$, we can obtain the PDD as shown in Fig. 2.14(b)

For $x' < -X$, $\qquad\qquad \dfrac{d}{dx} P(x \leqslant x') = 0$

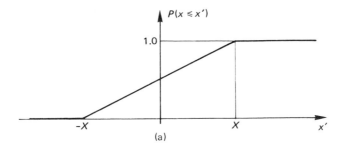

(a)

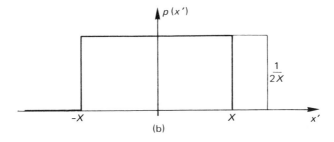

(b)

Fig. 2.14

$$\text{For } -X \leqslant x' \leqslant X, \qquad \frac{\mathrm{d}}{\mathrm{d}x}P(x \leqslant x') = \frac{1}{2X}$$

$$\text{For } x' > X, \qquad \frac{\mathrm{d}}{\mathrm{d}x}P(x \leqslant x') = 0.$$

The graph of PDD can be obtained from these. Note that the area under the PDD curve is unity.

Example

Find the PDD of a sine wave, Fig. 2.15. Consider the interval Δx at level x'. The probability that x will be found in the interval Δx at level x' will be

$$P(x' - \Delta x < x \leqslant x') = \frac{2\Delta t}{T}$$

But

$$\frac{\mathrm{d}x}{\mathrm{d}t} = \omega X \cos \omega t \simeq \frac{\Delta x}{\Delta t}$$

$$\therefore \ \Delta t \simeq \frac{\Delta x}{\omega X \cos \omega t}$$

Then

$$\frac{2\Delta t}{T} \simeq \frac{2\Delta x}{T\omega X \cos \omega t} = \frac{\Delta x}{\pi X \cos \omega t}.$$

Now $p(x')$ is the probability that x is in the interval $\mathrm{d}x$ at level x'. However,

$$p(x')\mathrm{d}x \simeq p(x')\Delta x$$

$$= \frac{\Delta x}{\pi X \cos \omega t}$$

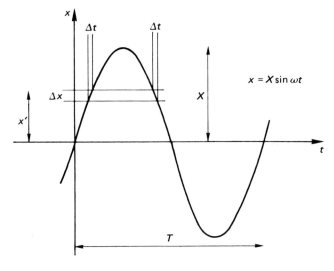

Fig. 2.15

$$p(x') = \frac{1}{\pi X \sqrt{1 - \sin^2 \omega t}}$$

$$= \frac{1}{\pi (X^2 - x^2)^{1/2}}.$$

Since x cannot be less than $-X$ or greater than X we must state this in the result, thus in general

$$p(x) = \frac{1}{\pi (X^2 - x^2)^{1/2}} \qquad \text{for } -X < x < X$$

$$= 0 \qquad\qquad\qquad \text{elsewhere.}$$

The CPD can be deduced from this using equation (2.2).

$$P(x \leqslant x') = \int_{-\infty}^{\infty} p(x)\mathrm{d}x.$$

Both the PDD and CPD are illustrated in Fig. 2.16. Observe that there is higher probability of finding x near to X and $-X$ than elsewhere, also that there is least probability of finding x near to zero. These are reasonable conclusions since x is changing slowly near X and $-X$ and rapidly near zero.

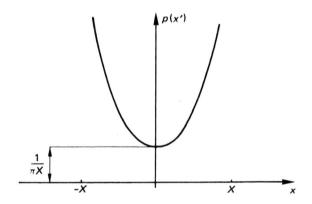

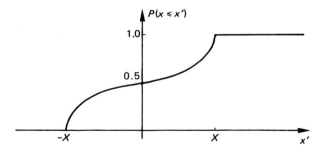

Fig. 2.16

2.9   **Discrete Probability Distributions**

Consider the square wave which is illustrated in Fig. 2.17. Being periodic we can obtain our information from a study of one period only as before. For levels chosen below $-X$, $P(x \leqslant -X) = 0$, for levels between $-X$ and X, $P(x \leqslant x') = 0.5$ and for levels above X, $P(x \leqslant X) = 1$. The corresponding CPD is shown in Fig. 2.17(a). If we apply equation (2.3)

$$p(x') = \frac{d}{dx} P(x \leqslant x')$$

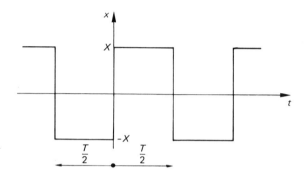

Fig. 2.17

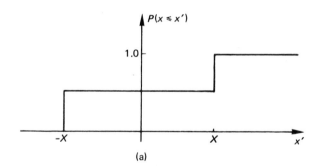

(a)

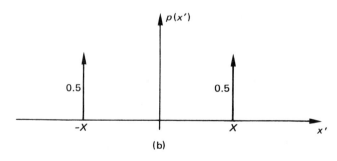

(b)

Fig. 2.18

we find that $p(x')$ consists of impulse functions (Fig. 2.18(b)) and can be expressed as

$$p(x') = \tfrac{1}{2}\delta(x' + X) + \tfrac{1}{2}\delta(x' - X).$$

Recall that the definition of an impulse requires

$$\int_{-\infty}^{\infty} \delta(x - a) = 1 \qquad (a \text{ is a constant}).$$

Hence in order to satisfy

$$\int_{-\infty}^{\infty} p(x)\,\mathrm{d}x = 1$$

in which $p(x')$ consists of two impulses we must have

$$\frac{1}{2}\int_{-\infty}^{\infty} \delta(x' + X)\,\mathrm{d}x + \frac{1}{2}\int_{-\infty}^{\infty} \delta(x' - X)\,\mathrm{d}x = 1$$

which confirms the above result.

PDD which are made up of impulses are known as discrete distributions, and Fig. 2.19 illustrates a typical example.

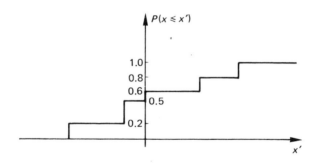

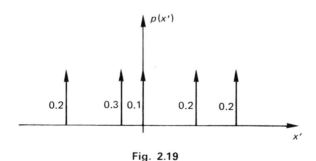

Fig. 2.19

2.10 Distributions and Averages for Discrete Signals

Re-writing equation (1.1) in expanded form we get

$$\bar{x} \simeq \frac{x_0}{N+1} + \frac{x_1}{N+1} + \ldots + \frac{x_N}{N+1}.$$

Now let us group the values of x as if we were constructing a histogram. Let the groups be x_a

to x_b, x_b to x_c etc. for which we imply that values observed at a boundary between intervals are included in the lower interval, that is, x_b is in the interval x_a to x_b, and x_c is in the interval x_b to x_c etc. We also assume that the intervals are small so that $x_b - x_a = x_c - x_b = \Delta x$ etc.

Let there be n_1 values in interval x_a to x_b
$\qquad\qquad n_2$ values in interval x_b to x_c etc.

Then we can write

$$\tilde{x} \simeq \frac{n_1}{N+1} \cdot x_b + \frac{n_2}{N+1} x_c + \dots$$

However, $\dfrac{n_1}{N+1}$ is approximately the probability that the signal is in the interval x_a to x_b,

$\dfrac{n_2}{N+1}$ that it is in the interval x_b to x_c etc.

$$\therefore \ \tilde{x} \simeq P(x_a < x \leqslant x_b)x_b + P(x_b < x \leqslant x_c)x_c + \dots$$

$$= \sum_{i=1}^{N} P(x_{i-1} < x \leqslant x_i)x_i \tag{2.6}$$

or

$$= \sum_{i=1}^{N} p(x_i)\Delta x \, x_i = \sum_{i=1}^{N} p(x_i)x_i\Delta x. \tag{2.7}$$

This gives a relationship between mean value and probability density.

If we use equation (1.1) on the data of Table 2.1 we get $\tilde{x} \simeq 99.78$. We can invoke equation (2.7) by using the information collected in section 2.2 and displayed by the histogram of Fig. 2.1.

$$\tilde{x} \simeq \qquad 0.22 \times 98.9 \times 1.0$$
$$+ 0.34 \times 99.9 \times 1.0$$
$$+ 0.32 \times 100.9 \times 1.0$$
$$+ 0.12 \times 101.9 \times 1.0$$
$$= \qquad 100.24.$$

The difference between the two results is due to the element of approximation in equation (2.7).

We can apply the same principles to the mean square value and obtain

$$\widetilde{x^2} \simeq \frac{n_1}{N+1}x_b^2 + \frac{n_2}{N+1}x_i^2 + \dots$$

and substituting probabilities as before we get

$$\widetilde{x^2} \simeq \sum_{i=1}^{N} P(x_{i-1} < x \leqslant x_i)x_i^2 \tag{2.8}$$

$$= \sum_{i=1}^{N} p(x_i)x_i^2 \Delta x. \tag{2.9}$$

Using equation (1.5) on our data we get:

$$\widetilde{x^2} \simeq \frac{1}{N+1}\sum_{i=0}^{N} x_i^2$$

giving

$$\widetilde{x^2} \simeq 9957.1$$

and

$$\text{r.m.s.} = \sqrt{9957.1} = 99.79.$$

Using equation (2.9) we get

$$\widetilde{x^2} \simeq 0.22 \times (98.9)^2 \times 1.0$$
$$+ 0.34 \times (99.9)^2 \times 1.0$$
$$+ 0.32 \times (100.9)^2 \times 1.0$$
$$+ 0.12 \times (101.9)^2 \times 1.0$$
$$= 10048.9$$

and

$$\text{r.m.s.} = \sqrt{10048.9} = 100.24.$$

Note that the computational labour involved in using equations (2.7) and (2.9) is less than that in using equations (1.1) and (1.5) but at the cost of previously finding relevant probabilities. Clearly equations (2.7) and (2.9) will give more accurate results when the intervals are smaller.

2.11 Distributions and Averages for Continuous Signals

In developing equation (2.7) we said

$$P(x_a < x \leqslant x_b) \simeq \frac{n_1}{N+1}$$

and so on for other values of x'. We could be more accurate if $x_b - x_a = \Delta x \rightarrow dx$ then

$$P(x' - dx < x \leqslant x') = p(x')dx$$

and on making the summation, which is now an integral, we get

$$\tilde{x} = \int_\alpha^\beta p(x) x \, dx.$$

The range of integration, α to β would have to be wide enough to include all possible values of the signal. To ensure this we make the range have infinite width:

$$\tilde{x} = \int_{-\infty}^\infty p(x) . x \, dx.$$

In exactly the same way we get

$$\widetilde{x^2} = \int_{-\infty}^\infty p(x) . x^2 \, dx.$$

Before any of these are of value for computational purposes we must know the density distribution, $p(x)$, as a function of x and then be clever enough at evaluating integrals.

Example

Find the mean value and mean square value of a saw tooth wave (Figs 1.14 and 2.12). We have deduced that the PDD of this wave is uniform between $-X$ and X at the value $\dfrac{1}{2X}$ and is zero outside this range, see Fig. 2.14(b). Thus we get

$$\tilde{x} = \int_{-\infty}^{\infty} p(x)x \,.\, \mathrm{d}x$$

$$= \int_{-X}^{X} \frac{1}{2X} x \,.\, \mathrm{d}x$$

$$= \frac{1}{2X}\left[\frac{x^2}{2}\right]_{-X}^{X}$$

$$= \frac{X^2 - X^2}{4X}$$

$$= 0.$$

which is a result we would expect. The mean square value will be

$$\widetilde{x^2} = \int_{-\infty}^{\infty} P(x) \,.\, x^2 \,\mathrm{d}x$$

$$= \int_{-X}^{X} \frac{1}{2X} x^2 \,\mathrm{d}x$$

$$= \frac{1}{2X}\left[\frac{x^3}{3}\right]_{-X}^{X}$$

$$= \frac{1}{6X}[X^3 + X^3]$$

$$= \frac{X^2}{3}.$$

This result is the same as that obtained in section 1.3.2 as expected.

2.12 Comments

We have now examined many ideas concerned with the amplitude characterisation of signals, the ideas have been developed mainly in the context of random signals but they are equally applicable to deterministic ones. It is reaffirmed that the best way of studying deterministic signals and their effect on and through systems is by exploiting their explicit methods of description. For random signals we can only use probablistic methods.

Whereas, in the realm of deterministic signals we can talk about sine waves, square waves, pulses and so on, when we are in the realm of random signals we may talk about Gaussian (or normal) signals, that is ones which have a Gaussian (or normal) PDD or a Rayleigh signal etc. (We can also talk about White Noise or Coloured Noise (signals) but the meaning of these terms will emerge later.) The CPD or the PDD of a signal is one of its statistical characteristics and remember, that inherent in this characteristic, is its mean value, mean square value and standard deviation.

If we performed observations of a signal and established its PDD and then repeated the experiment the next day and the day after and the day after that and so on, and obtained on each occasion the same results then the characteristic has not changed with time. We shall be examining other characteristics and if we find that these did not change with time either, then we could say that the signal was stationary (this term was mentioned in sections 1.2.7 and 1.2.8). Terms such as 'weakly stationary' and 'strongly stationary' will not be discussed here but they are discussed in advanced works on the subject. See, for example, Bendat and Piersol[1]

Signals whose characteristics do change with time are obviously non-stationary but unless otherwise stated all our studies will be relevant to stationary signals.

2.13   Multiple Signals and Ensemble Characteristics

In our work so far we have concerned ourselves with one signal only when discussing a particular concept. It could be that a system, natural or man-made, is subject to more than one signal simultaneously or that we are studying a lot of signals simultaneously. For example, in analysing ocean waves we may record them at several different geographical locations at the same time. In such a case we would obtain a set, or ensemble, of records as illustrated in Fig. 2.20.

In developing all the concepts so far we have obtained data about our signal by observing values of it at many points in time, in our case at equally spaced intervals. However, when we have a large ensemble it is possible to get our data points by measuring the values of each signal at one point in time. Referring to Fig. 2.20 we would obtain as data the values

$$_1x(t_1), \; _2x(t_1), \; _3x(t_1), \; _4x(t_1) \; \ldots \ldots$$

Using these values we could obtain a CPD and a PDD, a mean value, a mean square value and so on. Also we could manipulate the characteristics using relevant equations just as we did for those obtained from a single record.

The characteristics obtained in this way are called ensemble characteristics. Further, if we found that they were the same as those obtained from a single record we would say that the process, (i.e. the set of all signals), was ergodic. It can be shown that an ergodic process is stationary but that a stationary process is not necessarily ergodic.

2.14   Joint Probabilities

When we are considering the relationship between, say, two signals, or indeed, between one point on a signal and another point on the same signal, we need to know about joint probabilities.

Note that ensemble characteristics tell us about the ensemble and not about the relationships between individual members of the ensemble. We will now develop the essentials of joint probability for two signals (or two sets of data) but it is easy to extend the ideas, at least conceptually, to the joint probability for any number of signals. The main difficulties of extending the idea are, that of computational complexity and that of geometrical visualisation.

Suppose we are examining a manufacturing process consisting of two machines, one turning small rods to a specified nominal diameter and the other is boring bushes which are

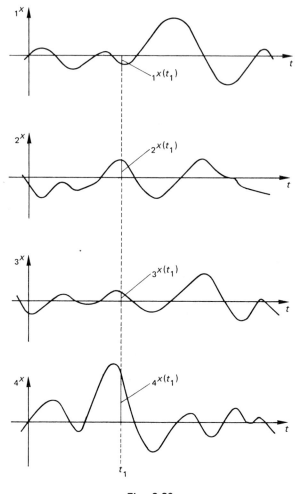

Fig. 2.20

intended to just slip onto the rods and rotate comfortably. Due to manufacturing variations there will be differences in size, both of the rods and of the bushes. Imagine selecting at random a rod and a bush, we will then find their fit to be either (a) as intended, (b) tight (or even no entry) or (c) slack. It is quite reasonable to ask ourselves the question, 'What is the probability of selecting a rod *and* a bush which fit as intended?' This is a joint probability, that is, a probability that two events happen together. In the world of signals we may want to know something such as 'What is the probability of observing a signal x at a value x' *and* observing a signal y at a value y'''? These ideas are expressed by means of joint probability distributions.

Consider the two signals illustrated in Fig. 2.21 and choose some value x' on x and some value y'' on y. The diagram is annotated to show:

(a)   The time intervals during which $x \leqslant x'$
(b)   The time intervals during which $y \leqslant y''$
(c)   The time intervals during which $x \leqslant x'$ *and* $y \leqslant y''$.

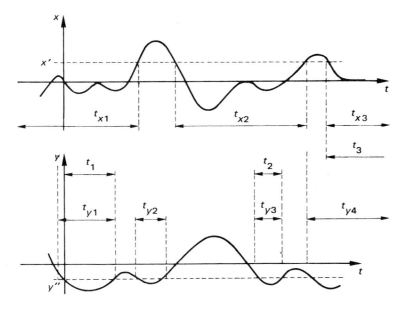

Fig. 2.21

It will be seen that the total time for the latter is $\tau = t_1 + t_2 + t_3$.
Using the ideas of section 2.6 we can say

$$P(x \leqslant x', y \leqslant y'') \simeq \frac{\tau}{T}$$

or more accurately

$$P(x \leqslant x', y \leqslant y'') = \underset{T \to \infty}{\text{Limit}} \frac{\tau}{T}.$$

Repeating for different values of x and y we can compute a large number of joint probabilities. If we choose 10 values for x and 10 for y the total number of probability values would be 100. It is also apparent that, in this case, probability is a function of two variables and can thus be plotted as a three-dimensional graph, Fig. 2.22, producing a surface instead of a curve as previously seen.

Note that this surface approaches a height of unity as both x' *and* y'' tend to large values. In the opposite direction the height of the surface approaches zero as both x' *and* y'' tend to large negative values. This surface would be expressed as a function of two variables x' and y'' such as $P(x \leqslant x', y \leqslant y'')$. This would be known as a bi-variate cumulative probability distribution.

Obviously we can produce a bi-variate probability density distribution which would also be a surface; a surface which would represent the slope of the CPD surface. The height of the PDD surface above the $x - y$ plane at some point (x', y'') would be $p(x', y'')$ obtained from

$$p(x', y'') = \frac{\partial^2}{\partial x \, \partial y} P(x \leqslant x', y \leqslant y''). \tag{2.10}$$

The inverse of this relationship is

$$P(x \leqslant x', y \leqslant y'') = \int_{-\infty}^{y''} \int_{-\infty}^{x'} p(x, y) \, \mathrm{d}x.\mathrm{d}y \tag{2.11}$$

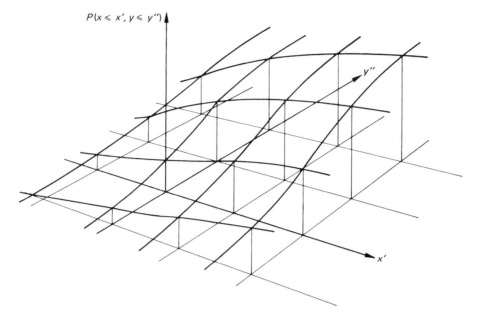

Fig. 2.22

from which we conclude that $p(x, y) \, dx \, dy$ is

$$P(x' - dx < x \leqslant x', \, y'' - dy < y \leqslant y'').$$

Recalling that the area of the strip above an element dx, namely $p(x) \, dx$ is $P(x' - dx < x \leqslant x')$ on the uni-variate PDD, it is clear that the volume over the elemental area $dx . dy$, namely $p(x, y) \, dx \, dy$ is $P(x' - dx < x \leqslant x', \, y'' - dy < y \leqslant y'')$. See Fig. 2.23. The joint Gaussian dis-

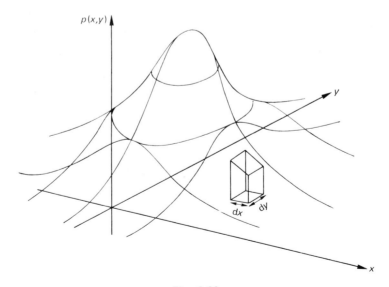

Fig. 2.23

tribution is

$$p(x, y) = \frac{1}{2\pi\sigma_1\sigma_2\sqrt{1-\rho_{12}^2}} \exp\left[\frac{-\left[\left(\frac{x-\mu_1}{\sigma_1}\right)^2 - 2\rho_{12}\left(\frac{x-\mu_1}{\sigma_1}\right)\left(\frac{y-\mu_2}{\sigma_2}\right) + \left(\frac{y-\mu_2}{\sigma_2}\right)\right]}{2\left(1-\rho_{12}^2\right)} \right]$$

where μ_1 and μ_2 are the respective mean values, σ_1 and σ_2 are the respective standard deviations and ρ_{12} is the correlation coefficient relating x and y. (See Chapter 3.)

If the two signals x and y are independent then it can be shown that

$$p(x, y) = p(x) \cdot p(y).$$

For the joint Gaussian distribution non-correlation means that $\rho_{12} = 0$ then

$$p(x, y) = \frac{1}{2\pi\sigma_1\sigma_2} \exp\left[\frac{-\left(\frac{x-\mu_1}{\sigma_1}\right)^2 - \left(\frac{y-\mu_2}{\sigma_2}\right)^2}{2} \right]$$

which is the same as

$$p(x) \cdot p(y) = \frac{1}{\sqrt{2\pi}\,\sigma_1} \exp\left[\frac{-\left(\frac{x-\mu_1}{\sigma_1}\right)^2}{2} \right] \frac{1}{\sqrt{2\pi}\,\sigma_2} \exp\left[\frac{-\left(\frac{y-\mu_2}{\sigma_2}\right)^2}{2} \right].$$

An example of a bi-variate process in which the two signals are independent is the one cited above where we were inspecting small rods and bushes. Clearly the sizes of the bores of the bushes are quite independent of the diameters of the rods since they are made on different machines. An example of a correlated process (bi-variate) is the input signal and the output signal of a system. The temperature of fluid emerging from a heat exchanger which correlates with the rate of energy input is an example of such a system.

Recall that the area under the PDD for a uni-variate curve is unity. This is the same as saying that the probability of the signal lying between $-\infty$ and $+\infty$ is unity. In the case of a joint distribution we have

$$\int_{-\infty}^{\infty} \int_{-\infty}^{\infty} p(x, y) \, dx \, dy = 1. \tag{2.12}$$

This is saying that the probability of finding x between $-\infty$ and $+\infty$ *and* y between $-\infty$ and $+\infty$ is unity! Geometrically it means that the 'total' volume under the surface is unity.

Example 1

A bi-variate distribution is given by

$$p(x, y) = e^{-(x+y)} \qquad \text{for } x \geqslant 0, y \geqslant 0$$

$$p(x, y) = 0 \qquad \text{elsewhere.}$$

What is the probability that $x \leqslant 3$ and $y \leqslant 1$? All we need to do is to evaluate

$$P(x \leqslant 3, y \leqslant 1) = \int_{-\infty}^{1} \int_{-\infty}^{3} P(x, y) \, dx \, dy.$$

However, since $P(x, y) = 0$ for $x < 0$ and $y < 0$ we can write

$$P(x \leqslant 3, y \leqslant 1) = \int_{0}^{1} \int_{0}^{3} e^{-(x+y)} \, dx \, dy$$

$$= \int_{0}^{1} [-e^{-(x+y)}]_{0}^{3} \, dy$$

$$= \int_{0}^{1} [-e^{-(3+y)} + e^{-y}] \, dy$$

$$= [e^{-(3+y)} - e^{-y}]_{0}^{1}$$

$$= (e^{-(3+1)} - e^{-1}) - (e^{-3} - 1)$$

$$= e^{-4} - e^{-1} - e^{-3} + 1$$

$$= 0.018 - 0.369 - 0.05 + 1$$

$$= 0.599.$$

Using the PDD of this example we can confirm equation (2.12)

$$\int_{0}^{\infty} \int_{0}^{\infty} e^{-(x+y)} \, dx \, dy$$

$$= \int_{0}^{\infty} [-e^{-(x+y)}]_{0}^{\infty} \, dy$$

$$= \int_{0}^{\infty} [0 + e^{-y}] \, dy$$

$$= [-e^{-y}]_{0}^{\infty} = 1.$$

Again using this PDD we can confirm the idea of independence

$$p(x, y) = e^{-(x+y)} = e^{-x} \cdot e^{-y}.$$

This can be written as

$$p_1(x) \cdot p_2(y) = e^{-x} \cdot e^{-y}$$

for which $p_1(x) = e^{-x}$ and $P_2(x) = e^{-y}$ which demonstrates independence.
Also note, using this result

$$P(x \leqslant 3) = \int_{0}^{3} e^{-x} \, dx$$

$$= [-e^{-x}]_{0}^{3}$$

$$= 0.95$$

and
$$P(y \leqslant 1) = \int_0^1 e^{-y} \, dy$$

$$= [-e^{-y}]_0^1$$

$$= 0.631$$

$$P(x \leqslant 3) . P(x \leqslant 1) = 0.95 \times 0.631 = 0.599$$

which is the same result as obtained previously.

2.15   The PDD for the Sum of Two Signals

Often two signals are added together, for example, when a wanted signal is transmitted through a channel, 'noise' is added so that the signal received is the sum of the wanted signal and the unwanted noise. It may well be necessary to find the characteristics of the sum; at present we are interested in the PDD of the sum. Let the two signals be x and y and let their sum be z, that is

$$z = x + y.$$

Assume we know $p(x, y)$, the joint PDD for x and y and we wish to find the PDD of z, namely $p_3(z)$. We approach this by finding

$$P(z \leqslant z').$$

Since $y = z - x$ for any value z, we get $y = z' - x$ for a particular value z'. Then values of $z \leqslant z'$ requires $y \leqslant z' - x$, therefore

$$P(z \leqslant z') = \int_{-\infty}^{\infty} \int_{-\infty}^{z'-x} p(x, y) \, dy . dx.$$

Now, using the relationship between $P(z \leqslant z')$ and $p_3(z)$, i.e. differentiation, we get

$$p_3(z') = \int_{-\infty}^{\infty} p(x, y = z' - x) \, dx. \tag{2.13}$$

If x and y are independent $P(x, y) = p_1(x) . p_2(y)$ giving

$$p_3(z') = \int_{-\infty}^{\infty} p_1(x) . p_2(z' - x) \, dx. \tag{2.14}$$

This is a convolution integral but its general significance will emerge later in a different context (Chapter 5).

Example

Find the PDD of the signal z for which $z = x + y$ and $p_1(x) = e^{-x}$ and $p_2(y) = e^{-y}$ as in the previous example.

From eqn. (2.14)
$$p_3(z) = \int_{-\infty}^{z} p_1(x) . p_2(z - x) \, dx$$

The reason for the upper limit being z and not ∞ will become clear when the convolution integral is studied in Chapter 5.

However, $p_1(x) = 0$ for $x < 0$ and $p_2(y) = 0$ for $y < 0$

thus

$$p_3(z) = \int_0^z p_1(x) \cdot p_2(z-x)\, dx$$

$$= \int_0^z e^{-x} e^{-(z-x)}\, dx$$

$$= \int e^{-z} \int_0^z e^{-x} e^x\, dx$$

$$= e^{-z} \int_0^z dx$$

$$= e^{-z} \cdot z.$$

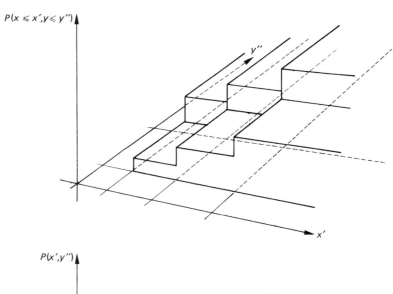

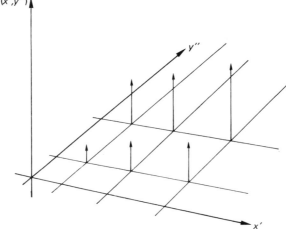

Fig. 2.24

2.16 Discrete Joint Distributions

We can define discrete joint distributions in a similar way as we did for a single variable. Such a distribution would appear as in Fig. 2.24.

2.17 Multi-variate Distributions

A natural extension of a bi-variate distribution is one involving three variables and theoretically we can go to n variables (n is, of course, any positive integer). In this case we have the following basic relationships

$$P(x_1, x_2, \ldots x_n) = \frac{\partial^n}{\partial x_1 \ldots \partial x_n} P(x_1 \leqslant x_1' \ldots x_n \leqslant x_n')$$

$$P(x_1 \leqslant x_1', \ldots x_n \leqslant x_n') = \int_{-\infty}^{x_n'} \int_{-\infty}^{x_{n-1}'} \ldots P(x_1, x_2 \ldots x_n) \, dx_1 \ldots dx_n.$$

A study of multi-variate distributions and their applications are beyond the scope of this work but they are mentioned here for completeness.

2.18 Summary

Since it is not possible to represent or model a random signal by means of an explicit mathematical function we have to resort to the ideas of probability. In doing this we find that the probability distribution, either the cumulative or the density, forms a convenient way of modelling the signal with regard to its instantaneous amplitude. We also find that its various average values are related to the distributions. These ideas can be applied to both discrete and continuous signals. We can also apply probabilities to more than one signal at once and represent a relationship between them by means of a joint probability distribution.

Exercises 2

1 For the voltage wave shown in Fig. 1.19 find the CPD and the PDD in the form of an annotated diagram. Using this result and the relevant relationships to find the mean value and the rms value.

2 Using the data in Question 2 of Chapter 1 find the CPD and the PDD in the form of bar charts or graphs. Use these charts to find the mean value and the rms value. Compare the results with those obtained in Question 2 of Chapter 1.

3 Find the value of k which makes p into a PDD.

$$
\begin{aligned}
p(x) &= kx && \text{for } 0 < x < 4 \\
&= k(8 - x) && \text{for } 4 \leqslant x \leqslant 8 \\
&= 0 && \text{elsewhere.}
\end{aligned}
$$

Plot this function together with the corresponding CPD. What is the probability that $x \leqslant 6$?

(Hint: Use the property $\int_{\infty}^{\infty} p(x) dx = 1$).

4 Determine which of the following are PDD's and thus find the corresponding mean values and mean square values.

(a) $p(x) = \dfrac{1}{b-a}$ for $a < x < b$

      $= 0$ elsewhere.

(b) $p(x) = \dfrac{1}{\pi} e^{-|x|}$ for $-3 < x < 1$

      $= 0$ elsewhere

(c) $p(x) = |x|$ for $-1 < x < 1$

      $= 0$ elsewhere.

5 Which of the following are discrete PDD's? Find the mean value and mean square value as appropriate.

(a) $p(x) = \dfrac{1}{12}$ for $x = 0$

      $= \dfrac{7}{12}$ for $x = 1$

      $= \dfrac{1}{4}$ for $x = 3$

      $= 0$ elsewhere.

(b) $p(x) = 2^{-|x|}$ for $x = \ldots -2, -1, 0, 1, 2, \ldots$

      $= 0$ elsewhere.

(c) $p(x) = \dfrac{2}{3}\left(\dfrac{1}{3}\right)^{x}$ for $x = 0, 1.2 \ldots$

      $= 0$ elsewhere.

6 A random signal x has an exponential PDD given by

$$p(x) = Ae^{-2|x|}$$

Find the mean value and the mean square value of x.

7 Two signals x and y have a joint PDD which is uniform, that is

$$p(x, y) = 1 \qquad \text{for } 0 < x < 1,\ 0 < y < 1$$
$$= 0 \qquad \text{elsewhere.}$$

Find the probability that

(a) $x < \frac{1}{2}$ and $y < \frac{1}{2}$

(b) $x > \frac{1}{3}$.

8 A random signal has a probability density distribution given by

$$p(x) = \frac{1}{2}\delta(x+2) + \frac{1}{4}\delta(x-3) + \frac{1}{4\sqrt{2\pi}}\exp\left(\frac{-x^2}{2}\right).$$

Find (a) The mean value, mean square value and variance
   (b) The probability that $x > 1$.
(Note and hint: for part (b) it will be necessary to refer to statistical tables showing areas under normal distributions).

3
Correlation

3.1 Introduction

So far we have been examining the amplitude properties of a signal but have made no reference whatsoever to the manner in which the signal behaves during time. If we look at the two signals illustrated in Fig. 3.1 we see at once that, whilst they are both random, they have quite different appearances, yet they could both have the same probability distributions. Obviously we must devise means whereby we can describe signals so as to differentiate between the characteristics clearly apparent in the figure. There are two ways and they are related to one another, one is by means of correlation and the other uses the idea of frequency. In this chapter we will look at correlation and its relevance to signals.

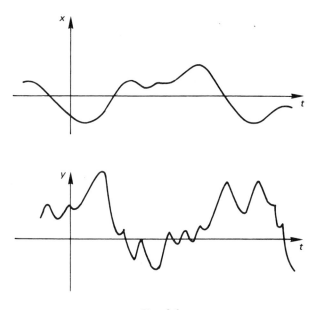

Fig. 3.1

3.2 Correlation Coefficient

Correlation means, of course, dependence, or the extent to which one thing depends on another. In terms of signals we say that correlation is a measure of the way in which one signal depends on another, or the way in which one part of a signal depends on another part of the same signal. Some examples of correlation are; (a) the pressure of steam in a boiler depends very much on the rate of fuel burning and; (b) in a sound wave the intensity at some instant in time is very dependent on the intensity of the previous moment.

In order to grasp the basic ideas of correlation let us look at the definition of correlation coefficient as considered in the study of probability and statistics. To develop the ideas let us consider a simple example. Imagine an extensive experiment to investigate the effectiveness of a new agricultural fertilizer. A large number of farmers co-operate in the experiment and each sows a given area of land with the same type of seeds. Each farmer then applies a different amount of fertiliser, recording the amount used. At the time of cropping each plot has its yield recorded so that at the end of the experiment there is a large number of records. Let the amounts of fertiliser used be

$$x_1, x_2, x_3, \ldots x_N$$

and the corresponding yields be

$$y_1, y_2, y_3, \ldots, y_N$$

where N is the number of farmers in the experiment. If we then plot all pairs of points on an $x-y$ domain we get a diagram which may look like one of those illustrated in Fig. 3.2. The

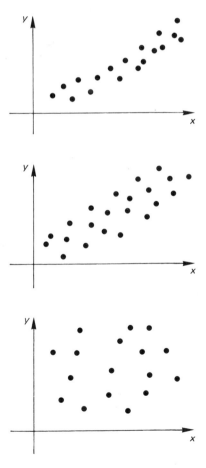

Fig. 3.2

first shows a high degree of correlation, the second shows a good degree of correlation, whilst the third shows no correlation.

In order to measure the degree of correlation we define the correlation coefficient as follows:

$$\rho_{xy} = \frac{(x_1 - \mu_x)(y_1 - \mu_y) + (x_2 - \mu_x)(y_2 - \mu_y) + \dots (x_N - \mu_x)(y_N - \mu_y)}{N\sigma_x\sigma_y}$$

$$= \sum_{i=1}^{N} \frac{(x_i - \mu_x)(y_i - \mu_y)}{N\sigma_x\sigma_y} \tag{3.1}$$

in which μ_x and μ_y are respective mean values and σ_x and σ_y are respective standard deviations.

Example

Ten students were examined in mathematics and engineering science and their results are shown in Table 3.1.

Table 3.1

Maths	78	36	98	25	75	82	90	62	65	39
Eng. Sc.	84	51	91	60	68	62	86	58	53	47

Is there any correlation between marks obtained in maths with those obtained in engineering science?

The mean value for maths is $\mu_x = 65$
The mean value for engineering science is $\mu_y = 66$
The respective standard deviations are $\sigma_x = 23.23$
$\sigma_y = 14.91$.

Table 3.2

$x_i - \mu_x$	$y_i - \mu_y$	$(x_i - \mu_x)(y_i - \mu_y)$
13	18	234
−29	−15	435
33	25	825
−40	−6	240
10	2	20
17	−4	−68
25	20	500
−3	−8	24
0	−13	0
−26	−19	494
		2704

$$\rho_{xy} = \frac{2704}{10 \times 23.23 \times 14.91}$$

$$= 0.78$$

We conclude that there is reasonable correlation in this case. There are other ways of performing the details of the calculation but these can be found in most books on statistics. Modern calculators have facilities for performing this type of calculation which allows one to avoid tedious manipulations.

3.3   Cross-Correlation Functions (Discrete Signals)

The basic correlation coefficient defined above is not suitable for quantifying correlation for signals. This is because it does not include the effect of delay and the inclusion of this effect is necessary because the dependence of one signal on another might be delayed. For example, if we increase the burning rate in a boiler it would be several seconds, or even minutes in a larger boiler, before the effect was felt at the point where steam pressure was measured. It is necessary to modify our correlation measure so as to include delay such that we can find the correlation at different values of delay. The result is that our correlation measure becomes a function of delay, hence the name Correlation Function. Delay can be negative which is of course advance, so to imply both we use the word 'shift'.

The basic definition is as follows but is only reasonably accurate when N is large.

$$R_{xy}[k] \simeq \frac{1}{N+1}[x_0 y_k + x_1 y_{k+1} + x_2 y_{k+2} + \ldots + x_N y_{N+k}]. \tag{3.2}$$

We can deduce the significance of this definition if we refer to Fig. 3.3.

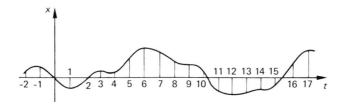

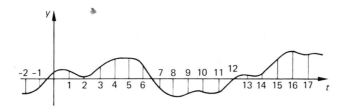

Fig. 3.3

Let us choose by way of an example, a shift k, of 3 intervals. We then calculate the cross-correlation function for this shift

$$R_{xy}[3] = \frac{1}{N+1}[x_0 y_3 + x_1 y_4 + \ldots x_N y_{N+3}].$$

To obtain the complete function we repeat this calculation for a range of values for k,

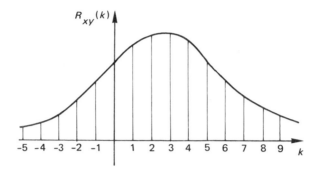

Fig. 3.4

including $k=0$, and obtain different values for the correlation function. These values can then be plotted with k as abscissa as shown in Fig. 3.4. Note that we can also make k negative, for example $k=-3$ gives,

$$R_{xy}[-3]=\frac{1}{N+1}[x_0y_{-3}+x_1y_{-2}+x_2y_{-1}+ \ldots . x_Ny_{N-3}].$$

Figure 3.4 indicates that maximum correlation occurs for a shift of 3 intervals approximately.

In using equation (3.2) it is important to realise that if we have $(N+1)$ values of x, that is

$$x_0, x_1, x_2, \ldots . . x_N$$

we need the $(N+1)$ relevant values of y as

$$y_k, y_{k+1}, y_{k+2}, \ldots . . y_{k+N}.$$

It follows that if we take successive values of k from 0 to k_{max} then the total number of values of y required will be $(N+1+k_{max})$. Often we only have the same number of y values as x values, in which case we need to modify equation (3.2) to

$$R_{xy}[k]\simeq\frac{1}{N+1-k}[x_0y_k+x_1y_{1+k}+ \ldots . x_{N-k}y_N]. \tag{3.3}$$

Modifying this for negative values of shift using the *same* data yields

$$R_{xy}[-k]\simeq\frac{1}{N+1-|k|}[x_Ny_{N-k}+x_{N-1}y_{N-1-k} \ldots . x_ky_0]. \tag{3.4}$$

Remember that we do not have y_{-1}, y_{-2} etc
   For example, if $N=10$ and $k=-3$ we get

$$R_{xy}[-3]\simeq\frac{1}{8}[x_{10}y_7+x_9y_6+x_8y_5+x_7y_4+x_6y_3+x_5y_2+x_4y_7+x_3y_0].$$

To illustrate the use of equations (3.3) and (3.4) we use the following data for which $N+1=15$.

x	10	11	15	8	5	9	26	30	32	20	22	27	25	0	6
y	10	9	12	13	14	18	7	8	11	26	28	29	16	19	23

$k=0$	$k=1$	$k=2$	$k=-2$
$x_i y_i$	$x_i y_{i+1}$	$x_i y_{i+2}$	$x_i y_{i-2}$
$10 \times 10 = 100$	$10 \times 9 = 90$	$10 \times 12 = 120$	$6 \times 16 = 96$
$11 \times 9 = 99$	$11 \times 12 = 132$	$11 \times 13 = 143$	$0 \times 29 = 0$
$15 \times 12 = 180$	$15 \times 13 = 195$	$15 \times 14 = 210$	$25 \times 28 = 700$
$8 \times 13 = 104$	$8 \times 14 = 112$	$8 \times 18 = 144$	$27 \times 26 = 702$
$5 \times 14 = 70$	$5 \times 18 = 90$	$5 \times 7 = 35$	$22 \times 11 = 242$
$9 \times 18 = 162$	$9 \times 7 = 63$	$9 \times 8 = 72$	$20 \times 8 = 160$
$26 \times 7 = 182$	$26 \times 8 = 208$	$26 \times 11 = 286$	$32 \times 7 = 224$
$30 \times 8 = 240$	$30 \times 11 = 330$	$30 \times 26 = 780$	$30 \times 18 = 540$
$32 \times 11 = 352$	$32 \times 26 = 832$	$32 \times 28 = 896$	$26 \times 14 = 364$
$20 \times 26 = 520$	$20 \times 28 = 560$	$20 \times 29 = 580$	$9 \times 13 = 117$
$22 \times 28 = 616$	$22 \times 29 = 638$	$22 \times 16 = 352$	$5 \times 12 = 60$
$27 \times 29 = 783$	$27 \times 16 = 432$	$27 \times 19 = 513$	$8 \times 9 = 72$
$25 \times 16 = 400$	$25 \times 19 = 475$	$25 \times 23 = 575$	$15 \times 10 = 150$
$0 \times 19 = 0$	$0 \times 23 = 0$	4706	3427
$6 \times 23 = 138$	4157		
3946			

$$R_{xy}[0] = \frac{3946}{15} = 263 \qquad R_{xy}[1] = \frac{4157}{14} = 296.9 \qquad R_{xy}[2] = \frac{4706}{13} = 362 \qquad R_{xy}[-2] = \frac{3427}{13} = 263.6$$

Figure 3.5 shows $R_{xy}[k]$ plotted as a function of k and it is easily seen that maximum correlation occurs for a shift of 3 intervals. This shift can be regarded as an advance of y or a delay of x and can be represented as follows.

		10	11	15	8	5	9	26	30		
		×	×	×	×	×	×	×	×		
		10	9	12	13	14	18	7	8	etc	$k=0$
		=	=	=	=	=	=	=	=		
		100	99	180	104	70	162	182	240		
		10	11	15	8	5	9	26	30		
		×	×	×	×	×	×	×	×		
	10	9	12	13	14	18	7	8	11	etc	$k=1$
		=	=	=	=	=	=	=	=		
		90	132	195	112	90	63	208	330		
		10	11	15	8	5	9	26	30		
		×	×	×	×	×	×	×	×		
10	9	12	13	14	18	7	8	11	26	etc	$k=2$
		=	=	=	=	=	=	=	=		
		120	143	210	144	35	12	286	780		
		10	11	15	8	5	9	26	30		
				×	×	×	×	×	×		
		10	9	12	13	14	18	etc	$k=-2$		
				=	=	=	=	=	=		
		150	72	60	117	364	540				

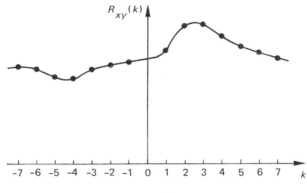

Fig. 3.5

Collecting the above expressions together we get

$$R_{xy}[k] \simeq \frac{1}{N+1-k} \sum_{i=0}^{N-k} x_i y_{i+k} \tag{3.5}$$

$$R_{xy}[-|k|] \simeq \frac{1}{N+1-|k|} \sum_{i=0}^{N-|k|} y_i x_{i+|k|}. \tag{3.6}$$

These expressions can be used for practical calculations but are approximate due to the need to use a finite number of values. However, for mathematical precision, as with probability, we must use the following for the purpose of definition.

$$R_{xy}[k] = \underset{N \to \infty}{\text{Limit}} \frac{1}{2N} \sum_{i=-N}^{N} x_i y_{i+k} \tag{3.7}$$

$$R_{xy}[-|k|] = \underset{N \to \infty}{\text{Limit}} \frac{1}{2N} \sum_{i=-N}^{N} y_i x_{i+|k|} \tag{3.8}$$

$$R_{xy}[-|k|] = \underset{N \to \infty}{\text{Limit}} \frac{1}{2N} \sum_{i=-N}^{N} x_i y_{i-|k|}. \tag{3.9}$$

Recall that as $N \to \infty$, $N+1-k \to N$ justifying the limits and the divisor $2N$. Note that with this notation y_0 and x_0 are in the centre of the range of data which makes the summation symmetrical and allows us to assimilate equations (3.8) and (3.9) into equation (3.7) and allows k to be both positive and negative.

3.4 Cross-Correlation Functions (Continuous Signals)

Naturally the concept of cross-correlation can be applied to continuous signals but, as before, we can invoke integration instead of summation and thus arrive at the definition of cross-correlation as

$$R_{xy}(\tau) = \underset{T \to \infty}{\text{Limit}} \frac{1}{2T} \int_{-T}^{T} x(t) \cdot y(t+\tau) \mathrm{d}t \tag{3.10}$$

where τ is a time shift, negative or positive. As with probability distributions, it is possible to evaluate cross-correlation functions for deterministic signals as the following example shows.

Example

Find the cross-correlation function relating x and y when

$$x = X \sin \omega t$$

$$y = Y \sin (\omega t + \phi).$$

Using the definition (3.10), we get

$$R_{xy}(\tau) = \operatorname*{Limit}_{T \to \infty} \frac{1}{2T} \int_{-T}^{T} X \sin \omega t \ Y \sin (\omega(t + \tau) + \phi) \mathrm{d}t.$$

For convenience in development of the argument write

$$R'_{xy}(\tau) = \int_{-T}^{T} \sin \omega t, \ \sin (\omega t + \phi') \mathrm{d}t$$

Nas - why did U do it?

in which $\phi' = \omega \tau + \phi$.

Then $\quad R_{xy}(\tau) = \operatorname*{Limit}_{T \to \infty} \frac{XY}{2T} R'_{xy}(\tau).$

Now, $\quad R'_{xy}(\tau) = \int_{-T}^{T} \sin \omega t (\sin \omega t \cos \phi' + \cos \omega t \sin \phi') \mathrm{d}t$

$$= \cos \phi' \int_{-T}^{T} \sin^2 \omega t \ \mathrm{d}t + \sin \phi' \int_{-T}^{T} \sin \omega t \cos \omega t \ \mathrm{d}t$$

$$= \frac{\cos \phi'}{2} \left[t - \frac{\sin 2\omega t}{2\omega} \right]_{-T}^{T} + \frac{\sin \phi'}{2} \left[-\frac{\cos 2\omega t}{2\omega} \right]_{-T}^{T}$$

$$= \frac{\cos \phi'}{2} \left[2T - \frac{\sin 2\omega T - \sin 2\omega(-T)}{2\omega} \right]$$

$$+ \frac{\sin \phi'}{2} \left[\frac{-\cos 2\omega T + \cos 2\omega(-T)}{2\omega} \right].$$

Then

$$R_{xy}(\tau) = \operatorname*{Limit}_{T \to \infty} \frac{1}{2T} XY \left[\frac{\cos \phi'}{2} \left[2T - \frac{\sin 2\omega T - \sin 2\omega(-T)}{2\omega} \right] \right.$$

$$\left. + \frac{\sin \phi'}{2} \left[\frac{-\cos 2\omega T + \cos 2\omega(-T)}{2\omega} \right] \right].$$

Dividing the term within braces by $2T$ and letting $T \to \infty$ gives

$$R_{xy}(\tau) = \tfrac{1}{2} XY \cos \phi'$$

$$= \tfrac{1}{2} XY \cos (\omega \tau + \phi).$$

This function is illustrated in Fig. 3.6.

Notice that maximum correlation occurs when $\tau = -\phi/\omega, \ -\phi/\omega \pm n. \ 1/f$ where $1/f$ is the period of one cycle and n is integer. This is physically reasonable since τ represents a shift of signal y and when this shift is equivalent to $-\phi/\omega$, x and y coincide.

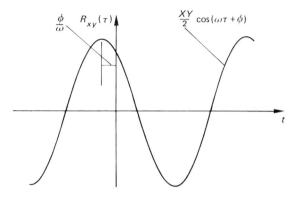

Fig. 3.6

3.5   Graphical Determination of Cross-Correlation Functions

It is very useful both from the point of view of computation and for the purpose of gaining insight, to consider a graphical approach to the determination of cross-correlation functions. To illustrate this, let us find the cross-correlation function for two square waves which have the same mark-space ratio and the same frequency but different amplitudes. Fig. 3.7 shows the pair together with relevant discrete values.

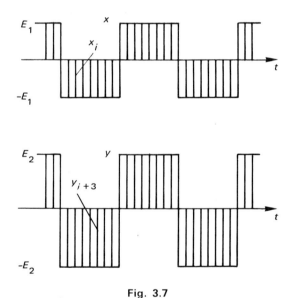

Fig. 3.7

   In principle we need to invoke equation (3.2) to effect the calculation and, taking a shift of three intervals by way of illustration, we would take terms such as x_i and y_{i+3}. This 'diagonal' multiplication, for this value of shift, could be converted to a 'vertical' multiplication by shifting y three intervals to the left as in Fig. 3.8. Re-numbering the ordinates allows us to take the products $x_0 y_0'$, $x_1 y_1' \ldots x_i y_i'$ etc., and construct a 'product' signal whose ordinates are $x_0 y_0'$, $x_1 y_1'$ etc.

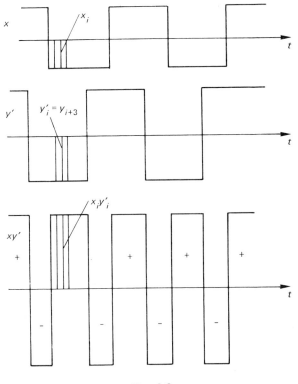

Fig. 3.8

Then
$$R_{xy}[3] \simeq \frac{1}{N+1-3}[x_0 y_3 + x_1 y_4 + \ldots]$$

$$= \frac{1}{N+1-3}[x_0 y'_0 + x_1 y'_1 + \ldots]$$

is proportional to the net area under the xy' curve. This operation can be repeated for different values of shift and a graph drawn showing net area plotted against shift. This will represent the required cross-correlation function, suitably scaled. Clearly different values of shift will produce different values of net area and it is obvious that maximum net area will occur when the shift is zero, that is, when the signals are in phase. Increasing the shift causes the area to decrease linearly to zero and then increase negatively to a negative maximum when the signals are in anti-phase. See Fig. 3.9. Note that the shift can be of opposite sign giving values for $R_{xy}[k]$ for negative values of k.

3.6   Correlation and Shift—Comment

The idea of shifting a signal can be used to gain an insight into the significance of correlation and what it can tell us. Consider the data used in the example in Section 3.3. Fig. 3.10 shows the data plotted as signals x and y. Fig. 3.11 shows the same data plotted but with y advanced by three intervals from which it is apparent that the signals are almost the same. Whilst this is easy to see in this example it is not always so, especially in cases where the signals are more complicated with a larger amount of data. Further, visual examination has little value for precise signal processing but, nevertheless, it can be of some value in

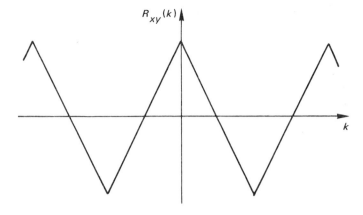

Fig. 3.9

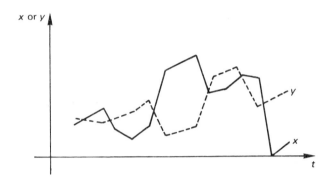

Fig. 3.10

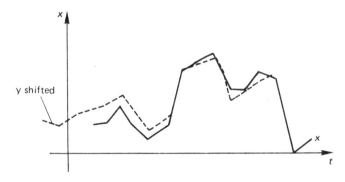

Fig. 3.11

obtaining a 'first look' to help in deciding processing strategy. Cross-correlation quantifies these ideas precisely. In passing it is worth noting, having found that there is a high correlation between x and y for an advance in y of three intervals, that x could be used to predict what value y might have in the future. Indeed formal methods are available for signal prediction.

3.7 Auto-Correlation Functions (Discrete Signals)

We have said that cross-correlation is a measure of the dependence of one signal on another, however, we can use the same concept to measure the dependence of one signal on itself! More specifically we use correlation to measure the dependence of the value of a signal at one instant in time with its value at another instant in time. For example, the value of a signal in the near future is likely to be very dependent on its value at the present. However, for random signals we would expect its value in the distant future to be negligibly dependent on its present value. Note, however, that we shall find that signals do exist whose value in the near future have little dependence on their present value. These turn out to be a common class of signal.

The measure of dependence is defined as the auto-correlation function as follows

$$R_{xx}[k] \simeq \frac{1}{N+1-k} [x_0 x_k + x_1 x_{1+k} + \ldots x_{N-k} x_N]$$

or

$$= \underset{N \to \infty}{\text{Limit}} \frac{1}{N} \sum_{i=0}^{N} x_i x_{i+k}$$

and in general

$$R_{xx}[k] = \underset{N \to \infty}{\text{Limit}} \frac{1}{2N} \sum_{i=-N}^{N} x_i x_{i+k}.$$

By way of example we will calculate the auto-correlation function for the signal x used in the example in Section 3.3.

Example

x,	10	11	15	8	5	9	26	30	32	20	22	27	25	0	6

$k=0$	$k=1$	$k=-1$
$x_i x_i$	$x_i x_{i+1}$	$x_{i+1} x_i$
$10 \times 10 = 100$	$10 \times 11 = 100$	$11 \times 10 = 110$
$11 \times 11 = 121$	$11 \times 15 = 165$	$15 \times 11 = 165$
$15 \times 15 = 225$	$15 \times 8 = 120$	$8 \times 15 = 120$
$8 \times 8 = 64$	$8 \times 5 = 40$	$5 \times 8 = 40$
$5 \times 5 = 25$	$5 \times 9 = 45$	$9 \times 5 = 45$
$9 \times 9 = 81$	$9 \times 26 = 234$	$26 \times 9 = 234$
$26 \times 26 = 676$	$26 \times 30 = 780$	$30 \times 26 = 780$
$30 \times 30 = 900$	$30 \times 32 = 960$	$32 \times 30 = 960$
$32 \times 32 = 1024$	$32 \times 20 = 640$	$20 \times 32 = 640$
$20 \times 20 = 400$	$20 \times 22 = 440$	$22 \times 20 = 440$
$22 \times 22 = 484$	$22 \times 27 = 594$	$27 \times 22 = 594$
$27 \times 27 = 729$	$27 \times 25 = 675$	$25 \times 27 = 675$
$25 \times 25 = 625$	$25 \times 0 = 0$	$0 \times 25 = 0$
$0 \times 0 = 0$	$0 \times 6 = 0$	$6 \times 0 = 0$
$6 \times 6 = 36$		
	4803	4803
5490		

$$R_{xx}[0] = \frac{5490}{15} = 366 \qquad R_{xx}[1] = \frac{4803}{14} = 343 \qquad R_{xx}[-1] = \frac{4803}{14} = 343$$

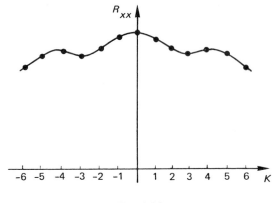

Fig. 3.12

If we plot these values of auto-correlation function against shift we get a curve as shown in Fig. 3.12.

Note that the value of $R_{xx}[k]$ is the same as $R_{xx}[-k]$; we could demonstrate here that auto-correlation is an even function but later we shall look at the properties of auto-correlation and find that, in general, it too is an even function.

3.8   Auto-Correlation Functions (Continuous Signals)

As with cross-correlation we can replace the summation by integration and obtain

$$R_{xx}(\tau) = \underset{T \to \infty}{\text{Limit}} \frac{1}{2T} \int_{-T}^{T} x(t) . x(t + \tau)\mathrm{d}t \tag{3.11}$$

in which the time shift τ can be positive or negative. Again, for deterministic signals we can evaluate $R_{xx}(\tau)$ and obtain an explicit function for it. The auto-correlation for a sine wave, for example, is easily found. Let $x = X \sin \omega t$, then

$$R_{xx}(\tau) = \underset{T \to \infty}{\text{Limit}} \frac{1}{2T} \int_{-T}^{T} X\sin \omega t \, X \, \sin \omega(t + \tau)\mathrm{d}t$$

We notice that this is the same as the example worked in section 3.4 except that $y = x$ and $\phi = 0$. Using the previous result with this substitution we get

$$R_{xx}(\tau) = \frac{1}{2}X^2\cos \omega \tau$$

which is illustrated in Fig. 3.13.

If there was a constant phase angle such that $x = X \sin (\omega t + \phi)$ it would be easy to show that ϕ would integrate out and give us the same result: $R_{xx}(\tau) = \frac{1}{2}X^2\cos \omega \tau$. The working is not difficult but is rather tedious and is left as an exercise. Note that this function is a cosine wave with the same frequency as the original.

$R_{xx}(\tau)$ for a square wave can be obtained using the results of section 3.5. If we make $y = x$, that is $E_1 = E_2 = E$ we get an auto-correlation function which is the same shape as that illustrated in Fig. 3.9 and the maximum values are the same as that for $R_{xx}(0)$. This is found

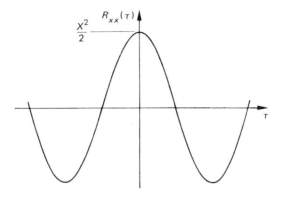

Fig. 3.13

as follows

$$R_{xx}(0) = \underset{T \to \infty}{\text{Limit}} \frac{1}{2T} \int_{-T}^{T} x^2(t)\,dt. \tag{3.12}$$

At all instants of time $x^2(t) = E^2$ thus

$$R_{xx}(0) = \underset{T \to \infty}{\text{Limit}} \frac{E^2}{2T} \int_{-T}^{T} dt$$

$$= E^2.$$

From both these examples we observe another property of auto-correlation functions. The definition of $R_{xx}(0)$ as stated in equation (3.12) is also the definition of the mean square value of a signal, hence $R_{xx}(0) = \overline{x^2}$.

3.9 Properties of Auto-Correlation Functions

The auto-correlation function has certain intrinsic properties which are useful when drawing conclusions about a signal from a study of its statistical properties. We will derive these properties in this section for both discrete and continuous signals as appropriate.

(i) Auto-correlation for zero delay is the same as the mean square value, i.e. $R_{xx}(0) = \widetilde{x^2}$
By definition

$$R_{xx}[0] = \underset{N \to \infty}{\text{Limit}} \frac{1}{2N} \sum_{i=-N}^{N} x_i x_i$$

$$= \underset{N \to \infty}{\text{Limit}} \frac{1}{2N} \sum_{i=-N}^{N} x_i^2.$$

This is the same as equation (1.6) except that in this case N is infinitely large and a double range for i is used. Thus

$$R_{xx}[0] = \widetilde{x^2}.$$

For continuous signals we obtained this result in section 3.8, i.e.

$$R_{xx}(0) = \overline{x^2}.$$

(ii) If the signal x has a non-zero mean value $\tilde{x}$, then its auto-correlation function will contain a constant term given by $(\tilde{x})^2$.

Let the signal x be decomposed into two components, one, the mean value $\tilde{x}$, and the other y a component with zero mean value such that $x = y + \tilde{x}$.

Then
$$R_{xx}[k] = \underset{N \to \infty}{\text{Limit}} \frac{1}{2N} \sum_{i=-N}^{N} (y_i + \tilde{x})(y_{i+k} + \tilde{x})$$

$$= L\sum (y_i y_{i+k} + y_i \tilde{x} + y_{i+k}\tilde{x} + (\tilde{x})^2)$$

where
$$L\sum \text{ means } \underset{N \to \infty}{\text{Limit}} \frac{1}{2N} \sum_{i=-N}^{N}$$

Then
$$R_{xx}[k] = L\sum y_i y_{i+k} + L\sum y_i \tilde{x} + L\sum y_{i+k}\tilde{x} + L\sum (\tilde{x})^2.$$

The first term on the right is $R_{yy}[k]$ by definition, the second and third terms are respectively

$$\tilde{x}L\sum y_i \quad \text{and} \quad \tilde{x}L\sum y_{i+k}$$

since $\tilde{x}$ is independent of the summation. However, $L\sum y_i = \tilde{y} = 0$ by definition, and $L\sum y_{i+k} = \tilde{y} = 0$ by definition. The last term is simply $(\tilde{x})^2$, hence

$$R_{xx}[k] = R_{yy}[k] + (\tilde{x})^2$$

For random signals $R_{yy}[k]$ approaches zero for large values of k, hence it is possible to estimate $\tilde{x}$ by examining $R_{xx}[k]$ at large values of k. The argument runs the same for continuous signals.

Let $x = y + \bar{x}$

$$R_{xx}(\tau) = \underset{T \to \infty}{\text{Limit}} \frac{1}{2T} \int_{-T}^{T} (y(t) + \bar{x}) \cdot (y(t + \tau) + \bar{x}) dt$$

$$= L\int (y(t) \cdot y(t + \tau) + y(t)\bar{x} + \bar{x}y(t + \tau) + (\bar{x})^2) dt$$

$$= L\int y(t) \cdot y(t + \tau) dt + L\int y(t) \cdot \bar{x} dt + L\int \bar{x}y(t + \tau) dt + L\int (\bar{x})^2 dt$$

$$= R_{yy}(\tau) + \bar{y}\bar{x} + \bar{x}\bar{y} + (\bar{x})^2$$

$$= R_{yy}(\tau) + (\bar{x})^2.$$

Note that $L\int$ means $\underset{T \to \infty}{\text{Limit}} \frac{1}{2T} \int_{-T}^{T}$.

(iii) Auto-correlation is an even function of k

that is
$$R_{xx}[-k] = R_{xx}[k].$$

Put
$$R_{xx}[k] \simeq \frac{1}{N+1-k} [x_0 x_k + x_1 x_{1+k} \ldots \ldots x_{N-k} x_N]$$

and
$$R_{xx}[-|k|] \simeq \frac{1}{N+1-|k|} [x_N x_{N-|k|} + x_{N-1} x_{N-1-|k|} + \ldots x_k x_0].$$

[Compare these equations with equations (3.4) and (3.5)]. Clearly $R_{xx}[k]$ and $R_{xx}[-k]$ are the same thus showing the even function property and is valid even when $N \to \infty$.

For continuous signals we have

$$R_{xx}(\tau) = L \int x(t) . x(t+\tau) dt$$

$$R_{xx}(-\tau) = L \int x(t) . x(t-\tau) dt.$$

Since we are dealing only with stationary signals we can shift the time origin by τ and not affect the result, hence $R_{xx}(-\tau)$ can be written as $R_{xx}(-\tau) = L \int x(t+\tau)x(t)dt$. This is the same as $R_{xx}(\tau)$ thus proving the even function property.

(iv) The value of an auto-correlation function is never greater than its value for zero delay, that is $R_{xx}[0] \geqslant R_{xx}[k]$.

Consider the following inequality

$$(x_i \pm x_{i+k})^2 \geqslant 0$$

that is, (the sum of two different values of $x)^2$ is greater than or equal to zero. Remember that we are concerned only with real values for x, thus squared terms are always positive. Expanding this we get

$$(x_i^2 \pm 2x_i x_{i+k} + x_{i+k}^2) \geqslant 0$$

or

$$x_i^2 + x_{i+k}^2 \geqslant \mp 2x_i x_{i+k}$$

Taking summations and limits we get

$$L\sum x_i^2 + L\sum x_{i+k}^2 \geqslant \mp L\sum 2x_i x_{i+k}.$$

The first term is, by definition, $\widetilde{x^2}$ and the second is also $\widetilde{x^2}$ since it is immaterial from where we take the count for i as the signals we are considering are stationary. The term on the right is, by definition, $2R_{xx}[k]$.

Hence

$$2\widetilde{x^2} \geqslant \mp 2R_{xx}[k]$$

i.e.

$$R_{xx}[0] \geqslant \pm R_{xx}[k] \text{ using property (i).}$$

Of course, the same is true for continuous signals

$$(x(t) \pm x(t+\tau))^2 = x^2(t) \pm 2x(t) \cdot x(t+\tau) + x^2(t+\tau) \geqslant 0$$

$$x^2(t) + x^2(t+\tau) \geqslant \mp 2x(t)x(t+\tau)$$

$$L\int x^2(t)dt + L\int x^2(t+\tau)dt \geqslant \mp L\int 2x(t) \cdot x(t+\tau)dt$$

i.e.

$$2\overline{x^2} \geqslant \mp 2R_{xx}(\tau)$$

or

$$R_{xx}(0) \geqslant \pm R_{xx}(\tau).$$

(v) If the signal x contains sinusoidal components then the auto-correlation function $R_{xx}[k]$ or $R_{xx}(\tau)$ contains corresponding sinusoidal components. The proof of this property is better demonstrated using continuous signals.

Let x contain both random and sinusoidal terms as follows, y is the random component.

$$x(t) = y(t) + X_1 \sin(\omega_1 t + \phi_1) + X_2 \sin(\omega_2 t + \phi_2) + \ldots$$

Then $x(t+\tau) = y(t+\tau) + X_1 \sin(\omega_1(t+\tau) + \phi_1) + X_2 \sin(\omega_2(t+\tau) + \phi_2) + \ldots$

By definition

$$R_{xx}(\tau) = L\int x(t) \cdot x(t+\tau)dt.$$

The integrand $x(t) . x(t+\tau)$ can be written as

$$x(t) \cdot x(t+\tau) = y(t) \cdot y(t+\tau)$$

$$+ y(t) \cdot X_1 \sin(\omega_1(t+\tau) + \phi_1)$$

$$+ y(t) \cdot X_2 \sin(\omega_2(t+\tau) + \phi_2)$$

$$+ \ldots.$$

$$+ X_1 \sin(\omega_1 t + \phi_1) \cdot y(t+\tau)$$

$$+ X_1^2 \sin(\omega_1 t + \phi_1) \cdot \sin(\omega_1(t+\tau) + \phi_1)$$

$$+ X_1 \sin(\omega_1 t + \phi_1) \cdot x_2 \sin(\omega_2(t+\tau) + \phi_2)$$

$$+ \ldots.$$

Taking the integral of each of these terms within the limits T and $-T$ as $T \to \infty$ and dividing by $2T$ $\left(\text{i.e. } \underset{T \to \infty}{\text{Limit}} \frac{1}{2T} \int_{-T}^{T} \right)$ shows that $R_{xx}(\tau)$ is made up of a number of components.

The first, $L\int y(t) \cdot y(t+\tau)dt$ is obviously $R_{yy}(\tau)$.

Also there will be terms of the form

$$L\int y(t) . X_i \sin(\omega_i(t+\tau) + \phi_i)dt$$

and

$$L\int X_i \sin(\omega_i t + \phi_i)y(t+\tau)dt.$$

Since y is random and $X_i \sin(\omega_i t + \phi_i)$ is periodic they are clearly uncorrelated, hence terms of this form will yield zero contribution. There will be terms of the form

$$L\int X_i^2 \sin(\omega_i t + \phi_i) . \sin(\omega_i(t+\tau) + \phi_i)dt.$$

Referring to Section 3.8 we deduce that this becomes

$$\tfrac{1}{2}X_i^2 \cos \omega_i \tau.$$

Finally, there will be terms of the form

$$L\int X_i \sin(\omega_i t + \phi_i)X_j \sin(\omega_j(t+\tau) + \phi_j)dt.$$

This is the average of the product of two sinusoids of different frequencies which is zero; a well known result in mathematical analysis. In the present context this means that the cross-correlation of two sinusoids of different frequencies is zero.

We conclude that

$$R_{xx}(\tau) = R_{yy}(\tau) + \tfrac{1}{2}X_1^2 \cos \omega_1 \tau + \tfrac{1}{2}X_2^2 \cos \omega_2 \tau + \ldots.$$

This means that the auto-correlation function of x contains the auto-correlation function of its random component plus a cosine term for each sinusoid present in x.

When using correlation as a method of signal analysis it is now clear that an examination of an auto-correlation function gives a lot of information about the signal. For example, it is readily seen whether or not a signal has a d.c. component or sinusoidal components. Indeed, visual examination of the signal itself may convey nothing as it could appear as a jumble with no identifiable components, yet an inspection of its auto-correlation function will readily identify periodic terms or a d.c. level. Fig. 3.14 illustrates this fact.

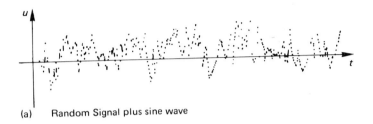

(a)   Random Signal plus sine wave

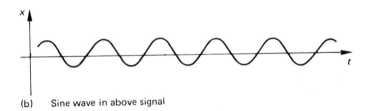

(b)   Sine wave in above signal

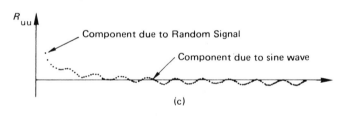

(c)

Fig. 3.14

3.10   Properties of Cross-Correlation

Cross-correlation functions also have properties which are useful but formal proofs are not given.

(i) Cross-correlation functions are not, in general even functions.

(ii) $R_{xy}[k] = R_{yx}[-k]$

For large values of N we can write

$$R_{xy}[k] \simeq \frac{1}{2N}[\ \ldots\ + x_{-2}y_{-2+k} + x_{-1}y_{-1+k} + x_0 y_k + x_1 y_{k+1} + \ \ldots\ ]$$

and

$$R_{yx}[-k] \simeq \frac{1}{2N}[\ldots + y_{-2}x_{-2-|k|} + y_{-1}x_{-1-|k|} + y_0 x_{0-|k|} + y_1 x_{1-|k|} + \ldots].$$

Setting, for example, $k=2$ we get

$$R_{xy}[k] \simeq \frac{1}{2N}[\ldots + x_{-2}y_0 + x_{-1}y_1 + x_0 y_2 + x_1 y_3 + \ldots]$$

and

$$R_{yx}[-k] \simeq \frac{1}{2N}[\ldots + y_{-2}x_{-4} + y_{-1}x_{-3} + y_0 x_{-2} + y_1 x_{-1} + \ldots].$$

Notice that the components of $R_{xy}[k]$ are the same as those of $R_{yx}[-k]$ so that as $N \to \infty$

$$R_{xy}[2] = R_{yx}[-2].$$

Similarly for any value of k, thus

$$R_{xy}[k] = R_{yx}[-k].$$

For continuous signals the argument runs as

$$R_{xy}(\tau) = L \int x(t) . y(t+\tau) dt$$

$$R_{yx}(-\tau) = L \int y(t) . x(t-\tau) dt.$$

Assuming, as with auto-correlation, that the signals are stationary a shift of time origin by τ will not affect the result, so we can write

$$R_{yx}(-\tau) = L \int y(t+\tau) x(t) dt$$

this is seen to be the same as $R_{xy}(\tau)$ hence

$$R_{xy}(\tau) = R_{yx}(-\tau).$$

(iii) The greatest value of the cross-correlation function is not necessarily at $k=0$ or $\tau=0$.

3.11 Auto-Correlation for the Sum of Two Signals

In our study of probability distributions we mentioned that signals can be added. We take this up again in the context of correlation.

Let

$$z = x \pm y.$$

Then

$$R_{zz}[k] = L \sum (x_i \pm y_i)(x_{i+k} \pm y_{i+k})$$

$$= L \sum (x_i x_{i+k} \pm x_i y_{i+k} \pm y_i x_{i+k} + y_i y_{i+k}).$$

On applying the operation $L \sum$ to each term we get

$$R_{zz}[k] = R_{xx}[k] \pm R_{xy}[k] \pm R_{yx}[k] + R_{yy}[k].$$

Exactly the same argument can be used for continuous signals to obtain

$$R_{zz}(\tau) = R_{xx}(\tau) \pm R_{xy}(\tau) \pm R_{yx}(\tau) + R_{yy}(\tau).$$

We conclude that the auto-correlation function of the sum of two signals is the sum of the auto-correlation functions of each signal plus the two cross-correlation functions. If x and y

are uncorrelated both $R_{xy}[k]$, $R_{yx}[k]$ and zero we get

$$R_{zz}[k] = R_{xx}[k] + R_{yy}[k].$$

Clearly these results can be extended to the case of the sum of more than two signals and if none of the components are correlated we would get for

$$z = x_1 + x_2 + x_3 + \ldots$$

$$R_{zz}[k] = R_{x_1 x_1}[k] + R_{x_2 x_2}[k] + R_{x_3 x_3}[k] + \ldots.$$

3.12   Ensemble Correlation Functions

In section 2.13 we considered probability distributions on a basis of ensemble characteristics. In a similar way we can consider correlation functions on the same basis. Refer to Fig. 3.15.

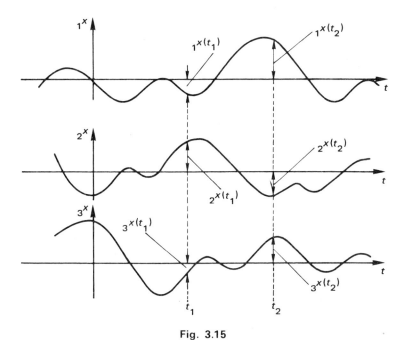

Fig. 3.15

We define the auto-correlation function as

$$R_{xx}[t_1, t_2] \simeq \sum_{i=1}^{N} {}_i x[t_1] \cdot {}_i x[t_2].$$

Note that the summation is now over the ensemble and $R_{xx}[t_1, t_2]$ can be evaluated for different values of t_1 and t_2. If the ensemble was ergodic and $t_2 = t + \tau$ we could then write

$$R_{xx}[\tau] \simeq \sum_{i=1}^{N} {}_i x[t] \cdot {}_i x[t + \tau].$$

In the sequel we will not be concerned with ensemble correlation, our principal interest lies in functions averaged over time. Ensemble functions are merely mentioned for completeness.

3.13   Other Forms of Correlation Functions

So far we have assumed in the discussion on correlation that the signals of interest are stationary, that is, they have existed from the infinite past with unchanged characteristics and will continue unchanged into the infinite future. We have applied the ideas to random signals and to periodic ones, however, for certain types of signal an alternative definition is necessary.

A definition appropriate for periodic signals is

$$R_{xx}[k] = \frac{1}{N+1} \sum_{i=0}^{N} x_i x_{i+k} \tag{3.13}$$

in which N is the number of intervals into which one period of the signal is divided, which will, of course, create $(N+1)$ ordinates. The corresponding form for continuous signals is

$$R_{xx}(\tau) = \frac{1}{T} \int_0^T x(t) . x(t+\tau) dt \tag{3.14}$$

in which T is the periodic time.

Since the signal is periodic we can divide by $(N+1)$ for discrete signals, or T for continuous signals, respectively, irrespective of the shift because we can always obtain values of x_{i+k}, or $x(t+\tau)$, from the next period if necessary. Fig. 3.16 shows this as well as the fact $x_{N+k} = x_k$ and $x_{N+i+k} = x_{i+k}$. To illustrate the use of equation (3.14) let us find the autocorrelation function of a sine wave, $x = X \sin \omega t$.

The integrand then becomes

$$X^2 \sin \omega t . \sin \omega(t+\tau) = X^2 \sin \omega t(\sin \omega t \cos \omega \tau + \cos \omega t \sin \omega \tau)$$

$$= X^2(\sin^2 \omega t \cos \omega \tau + \sin \omega t \cos \omega t \sin \omega \tau).$$

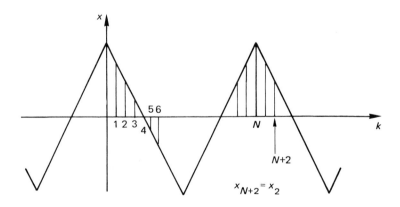

Fig. 3.16

Then

$$R_{xx}(\tau) = \frac{X^2}{T}\int_0^T \sin^2 \omega t \cos \omega \tau \, dt + \frac{X^2}{T}\int_0^T \sin \omega t \cos \omega t \sin \omega \tau \, dt$$

$$= \frac{X^2}{2T}\left[t - \frac{\sin 2\omega t}{2\omega}\right]_0^T \cos \omega \tau + \frac{X^2}{2T}\left[-\frac{\cos 2\omega t}{2\omega}\right]_0^T \sin \omega \tau$$

$$= \frac{X^2}{2T}\left[T - \frac{\sin 2\omega T}{2\omega}\right]\cos \omega \tau + \frac{X^2}{2T}\left[-\frac{\cos 2\omega T}{2\omega} + \frac{1}{2\omega}\right]\sin \omega \tau$$

$$= \frac{X^2}{2}\cos \omega \tau.$$

Since $\omega T = 2\pi$ giving $\sin 2\omega T = 0$ and $\cos 2\omega T = 1$. This gives the same result as previously but is computationally more convenient than equation (3.11) especially for machine computation.

We can define cross-correlation similarly:

$$R_{xy}(\tau) = \frac{1}{T}\int_0^T x(t)y(t+\tau)dt. \tag{3.15}$$

For an aperiodic signal we define correlation as

$$R_{xx}[k] = \underset{N \to \infty}{\text{Limit}} \sum_{i=N}^N x_i x_{i+k} \tag{3.16}$$

or

$$R_{xx}(\tau) = \int_{-\infty}^{\infty} x(t) \cdot x(t+\tau)dt. \tag{3.17}$$

(Note that in this case there is no division by a duration since it is infinite and the 'area' under an aperiodic signal is finite; if we did divide by infinity we would get zero for the auto-correlation function.) The use of equation (3.17) is ellucidated by the following example.

Example

Find the auto-correlation for the rectangular pulse shown in Fig. 3.17.
We need to consider three ranges of values for τ, two of which are illustrated in Fig. 3.18 and Fig. 3.19.

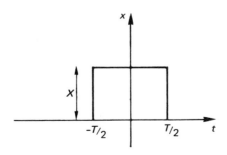

Fig. 3.17

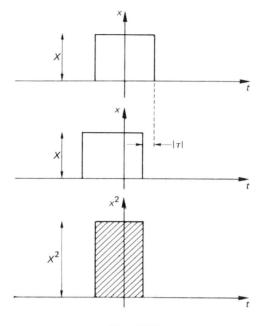

Fig. 3.18

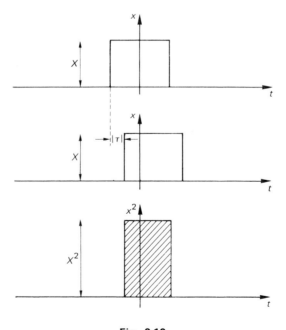

Fig. 3.19

(i) $0 \leqslant \tau \leqslant T$

$$R_{xx}(\tau) = \int_{-\frac{T}{2}}^{\frac{T}{2} - \tau} X^2 dt$$

$$= X^2 [t]_{-T/2}^{T/2 - \tau}$$

$$= X^2 \left[\frac{T}{2} - \tau + \frac{T}{2} \right]$$

$$= X^2 [T - \tau].$$

This is the shaded area shown in Fig. 3.18 and it clearly decreases linearly as $|\tau|$ increases.

(ii) $-T \leqslant \tau \leqslant 0$

$$R_{xx}(\tau) = \int_{-\frac{T}{2} + \tau}^{\frac{T}{2}} X^2 dt$$

$$= X^2 [T - \tau].$$

Since τ is negative in this case the area (Fig. 3.19) decreases as $|\tau|$ increases.

(iii) $|\tau| > T$.

In this case the delayed (or advanced) pulse is shifted so that the product $x(t) \, x(t + \tau)$ is always zero hence, for this range $R_{xx}(\tau) = 0$. Fig. 3.20 shows $R_{xx}(\tau)$.

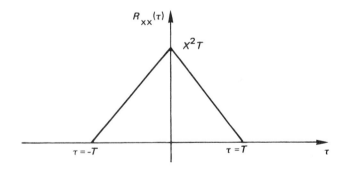

Fig. 3.20

Exercises 3

1   Find the auto-correlation function for the signals shown in Fig. 3.21 both by algebraic methods and by graphical methods.

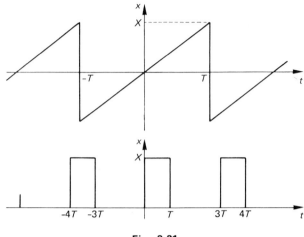

Fig. 3.21

2   Find the auto-correlation function for the signal whose data is listed in Question 2, Chapter 1, and for the signal data below the interval between each value being 1/20 sec. The best way is to write a short programme and input the data accordingly.

28	25	15	13	7	3	2	−5	−13	−16	−25	−22	−20	−16	−12	−8	−6	−2	−3	0
3	14	20	22	19	12	11	12	14	12	8	11	14	17	25	31	33	39	48	55
60	59	59	57	55	47	43	40	40	47	51	59	63	64	64	62	58	56	53	43
34	23	16	12	10	10	16	17	20	24	27	35	43	48	55	56	51	48	44	40
40	40	36	31	25	17	13	12	19	19	15	11	6	0	−3	−5	−5	−11	−16	−25
−26																			

3   Using the two sets of data in the previous problem find their cross-correlation functions, that is $R_{xy}(k)$ and $R_{yx}(k)$ for both negative and positive values of k. (A single programme could be written to compute both cross-correlation and auto-correlation.)

4   Find the cross-correlation function for the two signals shown in Fig. 3.22 both by algebraic and graphical methods.

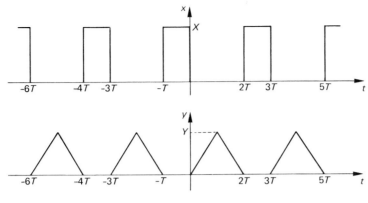

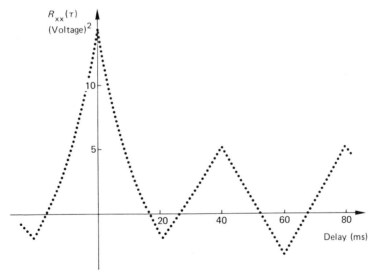

Fig. 3.22

5 Show that the cross-correlation function relating two periodic signals depends only on those harmonics that are present in both.

6 A digital correlator has produced correlogram as shown in Fig. 3.23. It purports to be an auto-correlogram of a 'certain signal', is the claim justified and what conclusions can be drawn about the signal?

Fig. 3.23

7 A simple mechanical structure is being shock tested by subjecting it to blows repeated at regular intervals of T seconds. Each blow can be approximated by a signal of the form $u(t) = Ue^{-at}$ and the resulting structural response at a point a short distance away from the point of impact by $y(t) = \dfrac{Y}{b-a} [e^{-at} - e^{-bt}]$ for each single blow.

If $T \gg \dfrac{1}{b} \gg \dfrac{1}{a}$ so that a particular response has decayed to negligible proportions before the next blow, find the cross-correlation function relating u with y and sketch its shape. If the response at a point further away from the point of impact was $y(t) = H(t-d)\dfrac{Y}{b-a}[e^{-a(t-d)} - e^{-b(t-d)}]$ where d is a time delay what effect would this have on the cross-correlation function?

8   A random signal has an auto-correlation function given by

$$R_{xx}(\tau) = 16e^{-2|\tau|}\cos 3.14\tau + 9.$$

Find the Mean value, Mean square value, the variance and the frequency of the periodic component.

Plot $R_{xx}(\tau)$ for the range $-4 \leqslant \tau \leqslant +4$ seconds.

4

Frequency Characterisation

4.1 Introduction

In the previous chapters the amplitude and time domain characteristics have been studied. The information extracted from the signals was embodied in the time histories which were observable and simply required some suitable mathematical operation on the time data. The interpretation of the extracted information sometimes required some care but the mathematical processes, particularly in discrete form, are relatively straightforward. When considering frequency characterisation, however, we have to invoke some mathematical procedures, which, to the uninitiated look somewhat cumbersome but often result in providing information which can be more readily interpreted than for the previous methods. It should be pointed out, however, that there is no more information available by looking at the frequency characteristics of a signal; it is just an alternative method of displaying the information.

The obvious question which must now be asked is: what kind of information can we expect from a frequency description of a signal? Referring to a sine wave (Fig. 4.1(a)) this is a single frequency function, which if a graph of frequency v signal amplitude were plotted, would result in Fig. 4.1(b).

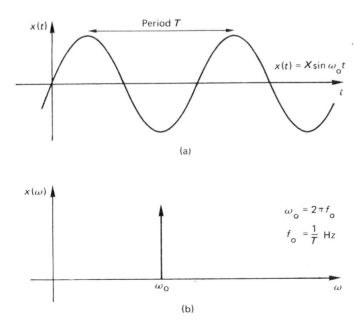

Fig. 4.1

This graph conveys as much information about the signal as does the time history, provided that the relationship between the two graphs is understood. If we try to extend this simplified approach to more complex signals, e.g. the random signals shown in Fig. 3.1, then obviously further information is required but we shall see that just as the correlation technique could differentiate between random signals so the frequency technique will provide similar information.

4.2   Integral Transforms

The basis of the frequency characterisation of signals is the Fourier transform, which is, of course, related to the Fourier series, as we shall see. Before continuing however, it is perhaps worth a few moments to put the concept of the transformation process into perspective. Consider the problem of multiplying two numbers together, e.g. 1693.36×0.306. To do this multiplication by hand would be rather tedious but we can transform these two numbers so that the multiplication process is replaced by addition, which is much easier and less time consuming. The transformation process in this instance is the use of a table of logarithms, and then, to obtain the required answer, the inverse or anti-logarithm tables are used, when the sum of the logs has been evaluated. Transformation techniques are therefore used to manipulate data or information into a more convenient form and the transformation process is always reversible.

One common transformation frequently used by engineers is the Laplace transform which is defined by the integral equation

$$F(s) = \int_0^\infty x(t) e^{-st} dt. \tag{4.1}$$

The time signal $x(t)$ is transformed to a signal in terms of the complex operator s ($s = \sigma + j\omega$). This rather daunting equation does of course greatly simplify the solution of linear differential equations by reducing them to algebraic equations in terms of s. Once again the transformation process is reversible so that time information can be retrieved. When the Laplace transform is used in the study of linear dynamic systems it is often convenient to investigate the response of a system to excitation by steady-state sine waves and in this case $s = j\omega$, i.e. $\sigma = 0$, and what we obtain is a frequency description of the system. The ideas of system characterisation will be investigated further in Chapters 5 and 6.

The transform of interest to the present work is the Fourier transform which is defined by equations (4.2) and (4.3)

$$X(j\omega) = \int_{-\infty}^\infty x(t) e^{-j\omega t} dt \tag{4.2}$$

$$x(t) = \frac{1}{2\pi} \int_{-\infty}^\infty X(j\omega) e^{j\omega t} dt. \tag{4.3}$$

The transformation from time to frequency equation (4.2) enables any time signal to be represented by a frequency component or set of frequency components depending on the nature of the signal. Some insight into the 'mechanics' of the transformation process can be gained by replacing the complex exponential $e^{-j\omega t}$ by the trigonometric equation

$$e^{-j\omega t} = \cos \omega t - j \sin \omega t$$

which can now be represented on a argand diagram as shown in Fig. 4.2. At some frequency

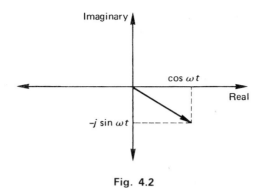

Fig. 4.2

ω_1, therefore, the vector will be at some fixed location and the frequency function $X(j\omega_1)$ is the summation of all the components of the signal at frequency ω_1. Referring back to the sine wave (Fig. 4.1) it is clear that the only frequency vector at which there will be any contribution from the signal is $\omega_0 (=2\pi f_0)$. For a random signal (Fig. 3.1), there will, however, be many frequencies at which the signal will have some component. As a further illustration of this idea consider the signal shown in Fig. 4.3 which is composed of the sum of two sinusoids. The two frequencies ω_1 and ω_2 quite clearly relate to the frequency components of the signal and the modulus of the amplitudes of the frequency components are in a corresponding ratio to the time amplitudes. You will observe that the frequency graph or spectrum contains both positive and negative frequencies. This can be explained by noting that the integration is over the limits from $-\infty$ to $+\infty$, this results in the mathematical concept of negative frequency which in reality does not exist.

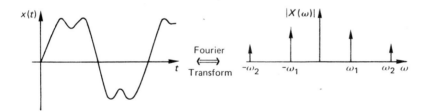

Fig. 4.3

In Fig. 4.3 only the modulus of the frequency components is shown. It is of course also possible to extract phase information from the signal as will be shown later.

Laplace transform ideas will not be developed further at this stage. However, we will return to the Laplace transform when studying the transmission of signals through linear systems in Chapter 5.

Before developing the Discrete Fourier Transform we will review the Fourier series in order to see how both the nature of the signal (periodic, non-periodic, random etc.) and the method of analysis determine the form of the spectrum, (i.e. continuous or discrete). First we will consider an important example.

Example

The frequency spectrum of a sinusoid, Fig. 4.1(b) was arrived at by intuition, now we will evaluate the Fourier transform of the sine wave. However, first we must consider the Fourier transform of delta functions.

$$\mathscr{F}\left[\delta(t-a)\right] = e^{-j\omega a}$$

and when $a = 0$

$$\mathscr{F}\left[\delta(t)\right] = 1.$$

Now by definition the Fourier transform is a reversible process.

Thus,   inverse $\mathscr{F}\left[e^{-j\omega a}\right] = \delta(t-a)$

i.e.

$$\frac{1}{2\pi} \int_{-\infty}^{\infty} e^{-j\omega a} e^{j\omega t} d\omega = \delta(t-a)$$

$$\frac{1}{2\pi} \int_{-\infty}^{\infty} e^{j\omega(t-a)} d\omega = \delta(t-a)$$

or

$$\int_{-\infty}^{\infty} e^{j\omega(t-a)} d\omega = 2\pi\delta(t-a)$$

again when $a = 0$

$$\int_{-\infty}^{\infty} e^{j\omega t} d\omega = 2\pi\delta(t)$$

and interchanging the variables gives:

$$\int_{-\infty}^{\infty} e^{j\omega t} dt = 2\pi\delta(\omega)$$

$$= 2\pi\delta(-\omega)$$

since the delta function is at $\omega = 0$.

We can now determine the Fourier transform of $\sin \omega_0 t$

$$\sin \omega_0 t = \frac{1}{2j}(e^{j\omega_0 t} - e^{-j\omega_0 t}).$$

Now

$$\mathscr{F}\left[e^{j\omega t}\right] = \int_{-\infty}^{\infty} e^{j\omega_0 t} e^{-j\omega t} dt$$

$$= \int_{-\infty}^{\infty} e^{j(\omega_0 - \omega)t} dt$$

and using the result above gives:

$$\mathscr{F}\left[e^{j\omega_0 t}\right] = 2\pi\delta(\omega_0 - \omega) = 2\pi\delta(\omega - \omega_0)$$

similarly

$$\mathscr{F}\left[e^{-j\omega_0 t}\right] = 2\pi\delta(\omega + \omega_0)$$

Hence
$$\mathscr{F}[\sin \omega_0 t] = \frac{1}{2j}[2\pi\delta(\omega - \omega_0) - 2\pi\delta(\omega + \omega_0)].$$

Cancelling terms and rationalising the complex denominator

$$\mathscr{F}[\sin \omega_0 t] = j\pi\delta(\omega + \omega_0) - j\pi\delta(\omega - \omega_0).$$

This confirms the result shown in Fig. 4.1(b) but note that the modulus of positive frequencies only, are shown there. The Fourier transform however, indicates that the negative frequency range also contains spectral information, i.e. $\delta(\omega + \omega_0)$. The concept of negative frequency is discussed later.

4.3   The Fourier Series and Periodic Signals

Most engineers and scientists are familiar with the application of the Fourier series to the analysis of periodic signals. The Fourier series of a periodic signal is simply the decomposition of the signal into its constituent sine and cosine terms at discrete frequencies. Consider the periodic signal shown in Fig. 4.4 which for convenience has the time datum fixed at the mid-point of one period.

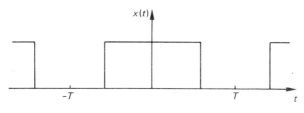

Fig. 4.4

The period of the signal is $2T$ and so the fundamental frequency component $f_0 = \frac{1}{2T}$ Hz or $\omega_0 = \pi/T$ rad s^{-1}. The Fourier series representation is then given by

$$x(t) = \frac{A_0}{2} + \sum_{N=1}^{\infty} (A_n\cos n\omega_0 t + B_n\sin n\omega_0 t). \tag{4.4}$$

The other frequency components which may be present in the signal are all multiples of the fundamental frequency and are called harmonics. The spectral coefficients A_n and B_n associated with each frequency component are given by

$$A_n = \frac{1}{T}\int_{-T}^{T} x(t)\cos n\omega_0 t\,\mathrm{d}t \tag{4.5}$$

where A_0 represents the d.c. or steady component of the signal (the mean value) and

$$B_n = \frac{1}{T}\int_{-T}^{T} x(t)\sin n\omega_0 t\,\mathrm{d}t. \tag{4.6}$$

Since the Fourier series separates the signal into sine and cosine terms this infers that at each frequency there will be both amplitude and phase information.

The amplitude of each component is

$$C_n = (A_n^2 + B_n^2)^{1/2}$$

and the phase of each component is

$$\phi_n = \tan^{-1} \frac{A_n}{B_n}.$$

The frequency information is normally presented as a plot of C_n versus ω and ϕ_n versus ω and since information is only available at discrete frequencies ($n\omega_0$, $n = 1, 2 \ldots \ldots \infty$) the familiar discrete spectrum is obtained (see Fig. 4.5). Alternatively, the spectrum could be displayed in terms of the coefficients A_n and B_n but this is not normally as readily interpreted as the amplitude and phase spectra.

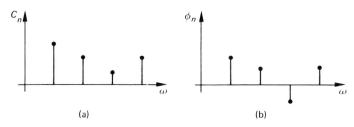

(a)   (b)

Fig. 4.5

In order to determine the Fourier coefficients the integral equations (4.5) and (4.6) must be evaluated. An alternative approach is to formulate the discrete equivalent of these equations and write a computer program to determine the coefficients. The discrete equivalent of equations (4.5) and (4.6) are:

$$A_n = \frac{2}{T} \sum_{k=0}^{N-1} x_k \cos\left(\frac{n2\pi k\Delta}{T}\right) . \Delta \tag{4.7}$$

$n = 0, 1, 2 \ldots$

$$B_n = \frac{2}{T} \sum_{k=0}^{N-1} x_k \sin\left(\frac{n2\pi k\Delta}{T}\right) . \Delta \tag{4.8}$$

The factor 2 is introduced to accommodate the change in observation period from $(-T \to T)$ to $(0 \to T)$ and where $\Delta =$ the time interval between samples. Now, $N\Delta = T$, N is the total number of samples available, thus, the coefficients become

$$A_n = \frac{2}{N} \sum_{k=0}^{N-1} x_k \cos(n2\pi k/N) \tag{4.9}$$

$n \geq 0$ (integer)

and

$$B_n = \frac{2}{N} \sum_{k=0}^{N-1} x_k \sin(n2\pi k/N) \tag{4.10}$$

clearly when $n = 0$

$$A_0 = \frac{2}{N} \sum_{k=0}^{N-1} x_k \qquad \text{the mean value.}$$

Example

Consider the square wave shown in Fig. 4.6.

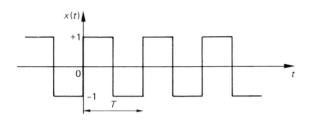

Fig. 4.6

The signal as defined is an odd function

i.e. $$x(-t)=-x(t).$$

This results in a Fourier series which contains only sine terms and therefore A_1, A_2, A_3 etc. are zero. Similarly, if $x(t)$ had been an even function [where $x(-t)=x(t)$] the resultant series would contain cosine terms only. In this particular case the signal is also symmetrical about the time axis and therefore $\bar{x}=0$, hence A_0 is also zero.

The coefficients of the sine terms B_n are now given by

$$B_n=\frac{2}{T}\int_{-0}^{T} x(t)\sin n\omega_0 t\,\mathrm{d}t.$$

In this example the signal is cyclic over the period T and we further note that the signal is not continuous over the whole period. Therefore the integral can be evaluated between the limits $t=0$ to $t=T/2$ when the magnitude of $x(t)$ is $+1$. The integral becomes

$$B_n=\frac{4}{T}\int_0^{T/2} 1.\sin n\omega_0 t\,\mathrm{d}t.$$

Having defined the fundamental period to be T, the fundamental frequency is given by

$$\omega_0=\frac{2\pi}{T}\qquad\left(\text{i.e.}f_0=\frac{1}{T}\text{Hz}\right).$$

Substituting for ω_0 yields

$$B_n=\frac{4}{T}\int_0^{T/2}\sin n\frac{2\pi}{T}.t\,\mathrm{d}t$$

$$=\frac{4}{T}\left[\frac{-T}{2\pi n}\cos n\frac{2\pi}{T}t\right]_0^{T/2}$$

$$=\frac{-4}{2\pi n}[\cos n\pi-1]$$

$$B_n=\frac{4}{\pi n}\qquad\text{when }n=1,3,5,7,\ldots.$$

$$B_n=0\qquad\text{when }n=0,2,4,6,\ldots.$$

Therefore the Fourier series for the square wave is given by

$$x(t) = \frac{4}{\pi}\left[\sin \omega_0 t + \frac{1}{3}\sin 3\omega_0 t + \frac{1}{5}\sin 5\omega_0 t + \ldots \frac{1}{n}\sin n\omega_0 t\right]$$

where n is an odd integer.

Now consider applying equation (4.10) to obtain the coefficients.

i.e.
$$B_n = \frac{2}{N}\sum_{k=0}^{N-1} x_k \sin\frac{n2\pi k}{N}$$

where $N =$ total number of samples in one cycle.
$x_k = +1$ $0 \leqslant k \leqslant N/2$
$x_k = -1$ $N/2 < k \leqslant N-1$
n is the coefficient to be evaluated.

B_n can be evaluated using a calculator but as this is extremely tedious a computer program should be written.

The results from such a program are given in Table 4.1 from which it can be seen that to attain reasonable accuracy a large number of samples (N) must be used.

Table 4.1

Analytical result		Numerical results			
		$N=40$	$N=80$	$N=160$	$N=320$
B_1	1.2732	1.2706	1.2725	1.2731	1.2732
B_3	0.4244	0.4165	0.4224	0.4239	0.4243
B_5	0.2546	0.2414	0.2514	0.2538	0.2544
B_7	0.1819	0.1632	0.1773	0.1807	0.1816
B_9	0.1415	0.1171	0.1355	0.1400	0.1411

4.4   Non Periodic Signals and the Discrete Fourier Transform (DFT)

Non periodic or aperiodic signals must be considered in a different way to periodic signals. Consider the signal shown in Fig. 4.7.

The signal is quite clearly non periodic, i.e. it is not repeated at regular time intervals. How can such a signal have a frequency content? The ideas of the Fourier series are now no

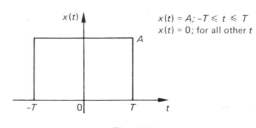

Fig. 4.7

longer adequate and so we must make use of the Fourier transform to try to explain the frequency characteristics of non periodic signals.

The Fourier transform is given by

$$X(j\omega) = \int_{-\infty}^{\infty} x(t)e^{-j\omega t}dt$$

$$= \int_{-T}^{T*} Ae^{-j\omega t}dt$$

$$= A\left[\frac{e^{-j\omega t}}{-j\omega}\right]_{-T}^{T}$$

$$= \frac{A}{-j\omega}[e^{-j\omega T} - e^{j\omega T}]$$

$$= \frac{A}{j\omega}[e^{j\omega T} - e^{-j\omega T}]$$

$$X(j\omega) = 2AT\frac{\sin \omega T}{\omega T}.$$

The result is a well known mathematical function $\frac{\sin x}{x}$ often called the sinc function. If $X(j\omega)$ is plotted against frequency ω then the result is as illustrated in Fig. 4.8. $X(j\omega)$ in this case is a continuous function of frequency (ω) and so wherever one chooses to observe a frequency then a component will exist. This is different from the case of the periodic signal which only has discrete frequency components which are related to the fundamental period of the signal.

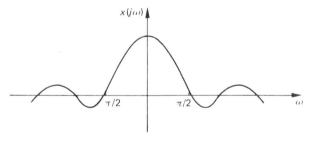

Fig. 4.8

The transition from discrete to continuous spectra can be explained heuristically by considering the aperiodic signal shown in Fig. 4.9.

This signal exists from $-\infty < t < +\infty$. The fundamental period of the signal can therefore be considered to be infinite and hence the fundamental frequency tends to zero (since $\omega_0 = 2\pi/T$), i.e. the discrete spectral lines merge to form a continuous spectrum.

The concept of the frequency content of a non periodic signal can now be extended to random signals; consider that shown in Fig. 4.10.

---

* Note that the limits of integration have been modified to suit the particular time function since $x(t) = 0$ outside the time interval $-T$ to $+T$.

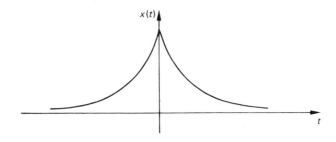

Fig. 4.9

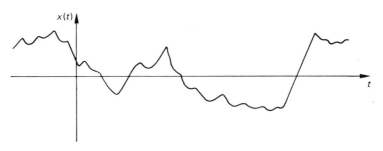

Fig. 4.10

Several assumptions have to be made about the signal, the most significant being that the signal is assumed to be stationary (Chapter 1). Provided that we have information about the signal from time $t = -\infty$ to $t = +\infty$ then we would expect to obtain a continuous spectrum. In practice, however, only a relatively short length of the signal would be available during some finite time interval, say $2T$ (see Fig. 4.11).

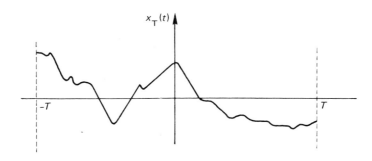

Fig. 4.11

The signal is now time limited and called $x_T(t)$ to differentiate it from the signal of infinite length. As T goes to infinity then the sample record becomes indistinguishable from the infinite record. By making the signal time limited, however, we have introduced a 'periodic' time, since in the absence of any further information the new time limited record is assumed to be periodic. The frequency spectrum obtained from this signal must now consist of discrete spectral lines located at the fundamental frequency ($\omega_0 = \pi/T$) and its harmonics.

Thus, the frequency content of the signal can be extracted from a Fourier series analysis of the signal.

We have now gone back to the original method of analysis but in the case of random signals it is not possible to evaluate the Fourier coefficients A_n and B_n from the integral equations (4.5) and (4.6), since $x(t)$ is not a deterministic signal an explicit mathematical function does not exist for $x(t)$.

In order to overcome this problem we must develop the discrete equivalent of the integral equation which can be shown to be related to the Fourier transform (Newlands). Indeed, the Fourier transform of a signal is the continuous equivalent of the Fourier coefficients expressed as a complex quantity, i.e.

$$X_n = A_n - jB_n$$

$$= \frac{1}{T}\int_0^T x(t)\cos\omega t\,dt - j\frac{1}{T}\int_0^T x(t)\sin\omega t\,dt$$
$$(\omega = n\omega_0) \tag{4.11}$$

$$= \frac{1}{T}\int_0^T x(t)[\cos\omega t - j\sin\omega t]\,dt$$

$$X_n = \frac{1}{T}\int_0^T x(t)e^{-j\omega t}\,dt. \tag{4.12}$$

Note that the range of integration has been changed and a factor of 2 has been taken outside the summation in equation (4.4) hence the Fourier series will require some manipulation to accommodate this. Equation (4.12) is now a time limited version of the Fourier transform equation (4.2). In order to be able to evaluate X_n equation (4.12) must be manipulated into discrete form so that the sampled values of the time signal, called a time series, can be used directly in the equation.

The instantaneous frequency ω is given by

$$\omega = 2\pi n/T \qquad [\text{See equations (4.7) to (4.10)}].$$

Some other parameters are redefined

$$dt = \Delta\text{—sample interval}$$
$$t = k\Delta \text{ —time elapsed from } t = 0$$
$$T = N \times \Delta\text{—total time (record length)}$$
$$x(t) \to x_k\text{—signal amplitude at sample interval } k.$$

Substituting these values into equation (4.12) and replacing the integral by a summation over the range $k = 0$ to $k = N - 1$ we have

$$X_n = \frac{1}{N\Delta}\sum_{k=0}^{N-1} x_k \exp\left[-j\frac{2\pi n}{T}.k\Delta\right]\Delta$$

and noting that $T = N\Delta$ this equation reduces to

$$X_n = \frac{1}{N}\sum_{k=0}^{N-1} x_k \exp\left[-j\frac{2\pi nk}{N}\right]. \tag{4.13}$$

Equation (4.13) is the discrete equivalent of the Fourier transform, the DFT, and gives the complex coefficient X_n corresponding to some fixed frequency ω_n. In order to compute a range of values of X_n, typically N coefficients would be required, then equation (4.13) must be evaluated N times, i.e.

$$X_0 = \frac{1}{N} \sum_{k=0}^{N-1} x_k \exp[0]$$

$$X_1 = \frac{1}{N} \sum_{k=0}^{N-1} x_k \exp[-j2\pi k/N]$$

$$X_{N-1} = \frac{1}{N} \sum_{k=0}^{N-1} x_k \exp[-j2\pi k(N-1)/N].$$

Once again we have spectral information at discrete frequencies just as for the Fourier series representation of deterministic signals.

Since the Fourier transform process is reversible the discrete equivalent of the inverse transform can be obtained by direct analogy with the continuous transform.

$$x_n = \sum_{k=0}^{N-1} X_k \exp\left[j\frac{2\pi nk}{N} \right]. \tag{4.14}$$

Having derived the discrete equivalent of the Fourier transform pair it is very tempting to write a computer program to evaluate X_n and the inverse x_n. Doing this, however, can lead to erroneous answers and result in very misleading conclusions being drawn about the characteristics for all but the most simple of signals. We will return to the problems of measuring and computing frequency spectra in Chapter 7.

4.5 Properties of the DFT

The transformation from time to frequency results in complex coefficients which are normally expressed in terms of the modulus and phase of the coefficients and then plotted as amplitude and phase spectra. Once again, as in the case of the continuous transform, the frequency range will be from $-\omega_{max}$ to $+\omega_{max}$ with only the positive frequency range being of practical interest.

At this stage it is worth noting some of the properties of the DFT

(i)   A time signal which is real and even will give frequency components which are real and even. (see Fig. 4.12).

(ii) A real odd signal will give frequency components which are imaginary and odd, e.g. a sine wave (Fig. 4.13).

(iii) An imaginary even signal results in imaginary and even frequency components.

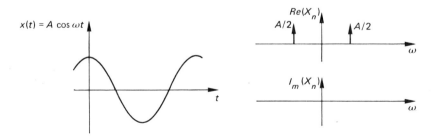

Fig. 4.12

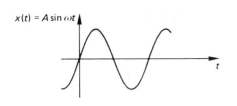

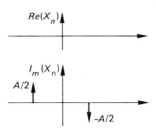

Fig. 4.13

(iv)   An imaginary odd signal results in real and odd frequency components.
(v)   Linearity.
   The Fourier transform is a linear transformation and therefore obeys the rule of superposition, i.e.

$$\mathscr{F}(x_n + y_n) = \int_{-\infty}^{\infty} x_n e^{-j\omega t}\,dt + \int_{-\infty}^{\infty} y_n e^{-j\omega t}\,dt$$

$$= \int_{-\infty}^{\infty} (x_n + y_n)e^{-j\omega t}\,dt = X_n + Y_n$$

or
$$x_n + y_n \Leftrightarrow X_n + Y_n$$

($\Leftrightarrow$ denotes the reversible Fourier transform process).

(vi)   Time shift
The effect of time shifting a time series can be realised in the frequency domain by multiplying the complex Fourier coefficients by a complex exponential term

$$x(n-i) \Leftrightarrow X_n e^{-j2\pi ki/N}.$$

This can be shown to be correct by taking the inverse transform

$$x(n-i) = \sum_{k=0}^{N-1} X_n e^{-j2\pi ki/N}\cdot e^{j2\pi nk/N}$$

$$x(n-i) = \sum_{k=0}^{N-1} X_n e^{j2\pi \frac{k}{N}(n-i)}$$

or in terms of the continuous transform

$$x(t-t_0) = \frac{1}{2\pi}\int_{-\infty}^{\infty} X(j\omega)e^{-j\omega(t-t_0)}\,d\omega$$

where the term $(n-i)$ in the complex exponential is directly related to the shift, i.e.

$$\text{time shift} = (n-i)\Delta \text{ seconds.}$$

(vii)   Frequency shift
The converse relationship to that for time shift holds for frequency shifting a signal by multiplying the time series x_n by the complex exponential.

$$x_n e^{+j2\pi ki/N} \Leftrightarrow X_{n-i}$$

To verify this apply the DFT to the modified time series

$$X_{n-i} = \frac{1}{N} \sum_{k=0}^{N-1} X_n \mathrm{e}^{+j2\pi ki/N} \cdot \mathrm{e}^{-j2\pi k_n/N}$$

$$= \frac{1}{N} \sum_{k=0}^{N-1} X_n \mathrm{e}^{-j2\pi \frac{k}{N}(n-i)}$$

which in terms of the continuous transform gives

$$X(\omega - \omega_0) = \int_{-\infty}^{\infty} x(t) \mathrm{e}^{-j(\omega-\omega_0)t} \, \mathrm{d}t.$$

(viii)   Convolution

The property of convolution will not be discussed here since it is related to the transmission of signals through linear systems and will therefore be discussed in detail in Chapter 6.

4.6   Spectral Density

The discussion so far has concentrated on the principles of the frequency characterisation of signals and has not been concerned with interpreting the results of frequency analysis. The reason for this is that most engineers do not require and indeed are often not concerned about raw Fourier transformed information. The normal method of displaying and interpreting spectral information is by means of 'power spectra' when the amplitude of the spectrum is defined in terms of Power Spectral Density (PSD).

The term power spectra, although a little misleading in some applications, arose from the early work done by electrical engineers in this field. This work was concerned with the power dissipated by an electrical component or circuit. In many engineering applications, however, the power associated with a system is proportional to the modulus of the frequency component squared.

i.e.
$$X_n = A_n + jB_n$$

$$|X_n|^2 = |A_n^2 + B_n^2|$$

or
$$\text{Power} \propto X_n^2.$$

The power density is then defined as the power per frequency interval or

$$\mathrm{PSD} = \frac{|X_n|^2}{\omega_0} \qquad \left[\frac{(\text{Amplitude})^2}{\text{rad s}^{-1}} \right] \tag{4.15}$$

where the amplitude will have the unit of the signal being investigated, e.g. $(\mathrm{m\,s}^{-2})^2$ in the case of acceleration. When the frequency is expressed in cyclic units (Hz) then the PSD becomes

$$\mathrm{PSD} = \frac{|X_n|^2}{f} \qquad \left[\frac{(\text{Amplitude})^2}{\text{Hz}} \right].$$

This is a rather simplistic approach to the idea of power spectral density but it does convey the basic concept of spectral density.

The derivation shown above works well for discrete spectra, but is not valid, however, for continuous spectra obtained from the Fourier integral equation. Indeed the concept of spectral density arose because of the restrictions imposed by the continuous Fourier transform. The continuous Fourier transform when applied to a signal of finite length T will yield real and imaginary components of the signal, the amplitudes of which are proportional to $1/T$ [equation (4.11)]. As T now becomes larger and the discrete spectral coefficients merge to yield a continuous spectrum (as $T\to\infty$), then the amplitude of the spectral coefficients will tend to zero. In order to overcome this problem we can divide by the frequency interval (which is tending to zero) and hence reach a limiting situation where the spectral density reaches a fixed value for a particular signal at some frequency.

The PSD must now be linked to the Fourier transform via mathematics. The outline of the proof is now given for those readers who wish to gain further insight into the concepts.

The power conveyed by a signal is determined by calculating the mean squared value of the signal which for the purposes of this discussion is assumed to be time limited.

$$\text{Mean squared value} = \frac{1}{2T}\int_{-T}^{T} x_T^2(t)\mathrm{d}t$$

Now
$$x(t) = \frac{1}{2\pi}\int_{-\infty}^{\infty} X(j\omega)e^{j\omega t}\mathrm{d}\omega \tag{4.16}$$

$$\therefore \int_{-\infty}^{\infty} x(t).x(t)\mathrm{d}t = \int_{-\infty}^{\infty} x(t).\frac{1}{2\pi}\int_{-\infty}^{\infty} X(j\omega)e^{j\omega t}\mathrm{d}\omega.\mathrm{d}t$$

$$= \frac{1}{2\pi}\int_{-\infty}^{\infty}\int_{-\infty}^{\infty} x(t)e^{j\omega t}\mathrm{d}t\,X(j\omega)\mathrm{d}\omega$$

$$= \frac{1}{2\pi}\int_{-\infty}^{\infty} X(-j\omega).X(j\omega)\mathrm{d}\omega \tag{4.17}$$

$$\therefore \int_{-\infty}^{\infty} x(t).x(t)\mathrm{d}t = \frac{1}{2\pi}\int_{-\infty}^{\infty} |X(j\omega)|^2\mathrm{d}\omega. \tag{41.8}$$

$$(X(-j\omega).X(j\omega)) = (|X(j\omega)|^2).$$

This result is known as Parsevals theorem and can be applied to the time limited signal $x_T(t)$

$$\int_{-T}^{T} x_T^2(t)\mathrm{d}t = \frac{1}{2\pi}\int_{-\infty}^{\infty} |X_T(j\omega)|^2\mathrm{d}\omega \tag{4.19}$$

dividing both sides by $2T$ gives

$$\frac{1}{2T}\int_{-T}^{T} x_T^2(t)\mathrm{d}t = \frac{1}{4\pi T}\int_{-\infty}^{\infty} |X_T(j\omega)|^2\mathrm{d}\omega \tag{4.20}$$

this can be re-written as

$$= \frac{1}{2\pi}\int_{-\infty}^{\infty}\left[\frac{|X_T(j\omega)|^2}{2T}\right]\mathrm{d}\omega \tag{4.21}$$

where $|X_T(j\omega)|^2/2T$ is the spectral density. Since the left hand side of the equation is the average 'power' conveyed by the signal the right hand side is the summation (integral) over the frequency range $-\infty$ to $+\infty$ of the spectral density, hence the spectral density is related to the power of the signal.

or
$$\frac{1}{2T}\int_{-T}^{T} x_T^2(t)\mathrm{d}t = \frac{1}{2\pi}\int_{-\infty}^{\infty} S_T(\omega)\mathrm{d}\omega. \tag{4.22}$$

where
$$S_T(\omega) = \frac{|X_T(j\omega)|^2}{2T}.$$

The area under the PSD curve is therefore equal to the mean squared response which in the limit becomes

$$S(\omega) = \lim_{T \to \infty} \frac{1}{2T} |X_T(j\omega)|^2. \tag{4.23}$$

Note that $S(\omega)$ is real valued and is an even function of frequency, see Fig. 4.14. The equation for PSD can be used for the estimation of power spectra by means of the DFT. The algorithm for estimating spectra will be discussed in detail in Chapter 7.

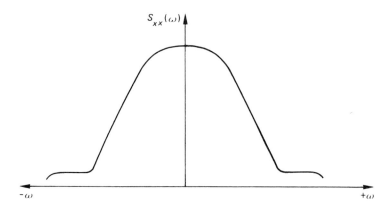

Fig. 4.14

The mathematical definition of PSD given in equation (4.15) is not precise and to obtain an accurate estimate of the PSD the individual estimates must be averaged in some way. The necessity for averaging and the methods of averaging which are commonly used are discussed in Chapter 7.

In order to further differentiate between the PSD of one signal with that of another it is common practice to use a double suffix notation as subscript, e.g. $S_{xx}(\omega)$ or $S_{yy}(\omega)$ which denote the PSD's of signals $x(t)$ and $y(t)$ respectively.

The typical spectrum shown in Fig. 4.14 has both positive and negative frequency content, the negative half of the spectrum being a mirror image of the positive half. The negative portion of the spectrum therefore contributes half the 'power' and so a positive frequency PSD function $G_{xx}(\omega)$ can be defined by

$$G_{xx}(\omega) = 2S_{xx}(\omega)$$

The PSD is now expressed in the form which is most familiar to engineers.

We are now in a position to look at typical spectra from random signals and to extract information from the signal in terms of frequency components.

4.7 Cross Spectra

The idea of one signal having some relationship with another signal has already been introduced in section 3.3. Here the relationship was quantified in terms of the cross-correlation function. In the frequency domain this relationship is quantified by the Cross Spectral Density (CSD). If the two signals are $x_T(t)$ and $y_T(t)$, time limited portions of the signals, then the Fourier transforms of the signals are

$$X_T(j\omega) = \int_{-T}^{T} x_T(t)e^{-j\omega t}dt \qquad (4.24)$$

$$Y_T(j\omega) = \int_{-T}^{T} y_T(t)e^{-j\omega t}dt. \qquad (4.25)$$

The cross spectral density is now defined by

$$S_{xyT}(j\omega) = \underset{T \to \infty}{\text{Limit}} \frac{1}{2T}[X_T^*(j\omega).Y_T(j\omega)] \qquad (4.26)$$

where $X_T^*(j\omega) = X_T(-j\omega)$ the complex conjugate of $X_T(j\omega)$. The cross spectral density was obtained by direct analogy with the derivation of the PSD of a signal. Note, however, that the order of the suffix is significant.

$$S_{yx}(\omega) = \underset{T \to \infty}{\text{Limit}} \frac{1}{2T}[X(j\omega).Y^*(j\omega)]$$

$$= \underset{T \to \infty}{\text{Limit}} \frac{1}{2T}[X^*(-j\omega).Y(-j\omega)]$$

$$= S_{xy}^*(\omega) \qquad (4.27)$$

Similarly $\qquad\qquad S_{xy}(\omega) = S_{yx}^*(\omega). \qquad (4.28)$

The importance of cross-spectra will become more apparent when we discuss the characterisation of the dynamics of systems and the ideas of system identification. For the present we are concerned with understanding the concepts and definitions.

4.8 The Relationships Between Power Spectra and Correlation Functions

In the introduction to this chapter it was emphasized that time domain and frequency domain information about a signal were merely a display of the same information in different forms. This can now be shown to be true by investigating the Fourier transform of the auto-correlation function of a time limited signal $x_T(t)$.

$$x_T(t) = x(t) \qquad -T < t < T$$

$$x_T(t) = 0 \qquad \text{elsewhere}$$

The auto-correlation function of $x(t)$ is defined by

$$R_{xx}(\tau, T) = \frac{1}{2T} \cdot \int_{-T}^{T} x(t)x(t+\tau)dt. \qquad (4.29)$$

The Fourier transform of $R_{xx}(\tau, T)$ is given by

$$\mathscr{F}[R_{xx}(\tau, T)] = \int_{-\infty}^{\infty} R_{xx}(\tau, T)e^{-j\omega\tau}d\tau. \qquad (4.30)$$

Substituting for $R_{xx}(\tau, T)$

$$\mathscr{F}[R_{xx}(\tau, T)] = \int_{-\infty}^{\infty} \frac{1}{2T} \int_{-T}^{T} x(t) \cdot x(t+\tau) e^{-j\omega\tau} dt d\tau. \tag{4.31}$$

Substituting $x_T(t)$ for $x(t)$ and changing the limits of integration from $\pm T$ to $\pm\infty$

$$\mathscr{F}[R_{xx}(\tau, T)] = \frac{1}{2T} \int_{-\infty}^{\infty} \int_{-\infty}^{\infty} x_T(t) \cdot x_T(t+\tau) e^{-j\omega(t+\tau)} \cdot e^{j\omega t} \cdot dt d\tau$$

$$= \frac{1}{2T} \int_{-\infty}^{\infty} \left[\int_{-\infty}^{\infty} x_T e^{j\omega t} dt \right] x_T(t+\tau) e^{-j\omega(t+\tau)} dt d\tau. \tag{4.32}$$

Let $t+\tau = \lambda$; then $d\tau = d\lambda$ and

$$\mathscr{F}[R_{xx}(\tau, T)] = \frac{1}{2T} \int_{-\infty}^{\infty} X_T(-j\omega) x_T(\lambda) e^{-j\omega\lambda} d\lambda \tag{4.33}$$

$$= \frac{1}{2T} X_T(-j\omega) X_T(j\omega)$$

$$= \frac{1}{2T} |X_T(j\omega)|^2$$

$$\mathscr{F}[R_{xx}(\tau, T)] = S_{xx_T}(\omega). \tag{4.34}$$

The more general result, when $T \to \infty$ can now be obtained by inspection, i.e.

$$\mathscr{F}[R_{xx}(\tau)] = \int_{-\infty}^{\infty} R_{xx}(\tau) e^{-j\omega\tau} d\tau = S_{xx}(\omega). \tag{4.35}$$

The auto-correlation function is therefore related to the PSD by Fourier integral relationships and they form a Fourier transform pair, i.e. the inverse relationship is given by

$$R_{xx}(\tau) = \frac{1}{2\pi} \int_{-\infty}^{\infty} S_{xx}(\omega) e^{j\omega\tau} d\omega. \tag{4.36}$$

In many text books on signal processing, PSD is often defined as the Fourier transform of the auto-correlation function. This approach was not adopted here because the significance of spectral density cannot be fully appreciated by this route.

The Fourier transform relationship can be manipulated to yield an interesting result

$$S_{xx}(\omega) = \int_{-\infty}^{\infty} R_{xx}(\tau) e^{-j\omega\tau} d\tau$$

$$= \int_{-\infty}^{\infty} R_{xx}(\tau) [\cos \omega\tau - j \sin \omega\tau] d\tau$$

$$= \int_{-\infty}^{\infty} R_{xx}(\tau) \cos \omega\tau d\tau - j \int_{-\infty}^{\infty} R_{xx}(\tau) \sin \omega\tau d\tau. \tag{4.37}$$

Now $R_{xx}(\tau)$ is an even function (see section 3.9) and $\sin \omega\tau$ is an odd function, therefore the second integral is zero. The first integral can be simplified by noting that $\cos \omega\tau$ is an even function, hence

$$S_{xx}(\omega) = 2 \int_0^\infty R_{xx}(\tau) \cos \omega \tau d\tau. \tag{4.38}$$

This result once again proves that $S_{xx}(\omega)$ is an even function. A similar result can be obtained for the inverse transform

$$R_{xx}(\tau) = \frac{1}{\pi} \int_0^\infty S_{xx}(\omega) \cos \omega \tau d\omega. \tag{4.39}$$

These two expressions are known as the Wiener–Kinchine relations.

The corresponding results relating cross spectral density with the coss-correlation function are

$$R_{xy}(\tau) = \frac{1}{2\pi} \int_{-\infty}^\infty S_{xy}(\omega) e^{j\omega\tau} d\omega \tag{4.40}$$

$$S_{xy}(\omega) = \int_{-\infty}^\infty R_{xy}(\tau) e^{-j\omega\tau} d\tau. \tag{4.41}$$

The ideas of spectral density have been derived in terms of continuous time. The relationships given above all have their discrete versions which require an understanding of the concept of convolution before the complications can be fully explained. Any further discussion about discrete spectra will therefore be delayed until Chapter 7.

4.9   Properties of Spectral Density

(i)   Power spectral density (PSD) is a measure of the power per unit bandwidth of a signal, phase information relating the various components is therefore lost in this description of the signal.

(ii)   PSD is real valued since both power and frequency are real

(iii)   By definition PSD is non negative at all frequencies. In terms of electrical power in a resistor, a negative power would infer that power is generated in a resistor. This is clearly not possible.

(iv)   PSD is an even function of frequency—this has been proved in section 4.8

(v)   If the random signal contains sinusoidal components at $\omega_1, \omega_2, \omega_3 \ldots \omega_N$ then $S_{xx}(\omega)$ will contain impulses at $\omega_1, \omega_2 \ldots \omega_N$ and at corresponding points in negative frequency $-\omega_1, -\omega_2, \ldots -\omega_N$.

This property can be proved by using the Weiner–Kinchine relationship applied to property (v) of the auto-correlation function viz.

$$R_{xx}(\tau) = R_{yy}(\tau) + \frac{1}{2} \sum_{n=1}^N X_n^2 \cos \omega_n \tau.$$

This is the auto-correlation function for a random signal $y(t)$ plus harmonic components at frequencies $\omega_1 \ldots \omega_n$. The PSD is given by the Fourier transform of $R_{xx}(\tau)$

$$S_{xx}(\omega) = \mathscr{F}[R_{yy}(\tau)] + \mathscr{F}\left[\frac{1}{2} \sum_{n=1}^N X_n^2 \cos \omega_n \tau \right]$$

$$S_{xx}(\omega) = S_{yy}(\omega) + \frac{1}{2}X_n^2\pi\left[\sum_{n=1}^{N}(\delta(\omega-\omega_n)+\delta(\omega+\omega_n))\right]$$

(see the table of Fourier transform pairs to verify the Fourier transform of $\cos\omega\tau$).

$S_{xx}(\omega)$ is therefore composed of the spectrum of the random signal with discrete spectral components at the frequencies of the harmonic components. As in the case of the auto-correlation function this provides a means of extracting a signal from corrupting noise. A typical example is the PSD of the time signal shown in Fig. 3.14 (a) which looks like a random signal but in fact the signal is dominated by a single frequency sine wave when the PSD is measured. See Fig. 4.15.

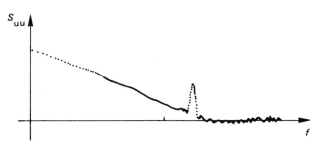

Fig. 4.15

Cross Spectral Density (CSD)

(i)   The cross spectral density function is a complex function of frequency and can therefore be expressed in terms of amplitude and phase.

(ii) $S_{xy}(\omega) = S_{yx}^*(\omega)$.

This can be proved by using the result in section 3.10

$$R_{xy}(\tau) = R_{yx}(-\tau)$$

and substituting for $R_{xy}(\tau)$ in equation (4.41)

i.e.

$$S_{xy}(\omega) = \int_{-\infty}^{\infty} R_{yx}(-\tau)e^{-j\omega\tau}d\tau.$$

Now let $\eta = -\tau$
Therefore

$$S_{xy}(\omega) = \int_{-\infty}^{\infty} R_{yx}(\eta)e^{j\omega\eta}(-d\eta)$$

$$= \int_{-\infty}^{\infty} R_{yx}(\eta)e^{j\omega\eta}d\eta$$

The term on the right-hand side is the complex conjugate of S_{yx}. This results from the positive exponential term in the transform.

Hence

$$S_{xy}(\omega) = S_{yx}^*(\omega)$$

and conversely

$$S_{yx}(\omega) = S_{xy}^*(\omega).$$

4.10 The Units of Spectral Density

$$S_{xx}(\omega) = \underset{T \to \infty}{\text{Limit}} \frac{1}{2T} |X(j\omega)|^2 \qquad (4.42)$$

From this definition the units of spectral density are. $(x^2)/$(unit angular frequency), where x is the unit of the time signal, e.g. volts, $m\,s^{-1}$, $kg\,s^{-1}$ etc. The units of cross spectral density are therefore given by $(xy)/$(unit angular frequency).

We have already seen that it is normal practice to consider only the positive frequency range of the PSD, and so we normally consider only the positive half of the cross spectral density and double the amplitude to maintain the same amplitude contribution.

In practice frequency is normally measured in Hz rather than rad/s^{-1}. In order to convert from (spectral density) in $rad\,s^{-1}$ to spectral density in Hz the PSD or CSD must be multiplied by 2π. Therefore to convert a PSD function $S_{xx}(\omega)$ to a single sided PSD as a function of $f(\text{Hz})$, $G_{xx}(f)$, the multiplying constant is 4π i.e.

$$G_{xx}(f) = 4\pi \times S_{xx}(\omega) \qquad (4.43)$$

and

$$G_{xy}(f) = 4\pi \times S_{xy}(\omega).$$

Exercises 4

1 Show that the Fourier series for the periodic wave form shown in Fig. 4.16 is given by

$$f(t) = \frac{1}{2} + \frac{2}{\pi} \sum_{n=0}^{\infty} \frac{1}{2n+1} \sin{(2n+1)}\,\omega t.$$

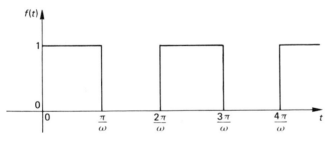

Fig. 4.16

2 Sketch the wave-form which results when the time function $f_n(t)$ is defined by

$$f_n(t) = f(t) \times \sin{\omega t}$$

where $f(t)$ is the Fourier series in the previous question. Hence show that the Fourier series of $f_n(t)$ can be written as

$$f_n(t) = \frac{1}{\pi} + \frac{1}{2} \sin{\omega t} - \frac{2}{3\pi} \cos{2\omega t} - \frac{2}{15\pi} \cos{4\omega t} + \ldots$$

Remember the trigonometric identities

$$\sin^2 x = \tfrac{1}{2}(1 - \cos{2x})$$

$$\sin{x} \sin{y} = \tfrac{1}{2}[\cos{(x-y)} - \cos{(x+y)}]$$

3   Determine the Fourier transform for the time function shown in Fig. 4.17.

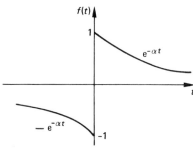

Fig. 4.17

4   Determine the Fourier transform for the time function shown in Fig. 4.18.

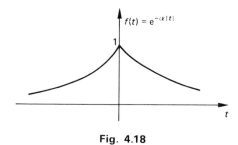

Fig. 4.18

5   With reference to properties 1 and 2 of the Discrete Fourier Transform, (the continuous transform has similar properties), explain the form of the frequency functions obtained in questions 3 and 4.

Add the time functions Fig. 4.17 and Fig. 4.18 and obtain the Fourier transform of the resulting time function. Could this result have been predicted from the Fourier transforms of the component time functions?

6   Use the DFT (equation 4.13) to determine the discrete spectra for the following time series.

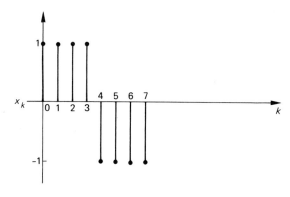

Fig. 4.19

Plot the following discrete spectra

$$Re(X_n) \, v \, n; \quad Im(X_n) v \, n; \quad |X_n| \, v \, n.$$

7 The autocorrelation function of a signal is shown in Fig. 4.20. Sketch the power spectral density of the signal.

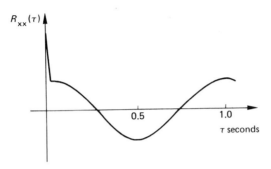

Fig. 4.20

8 The spectral density of a signal is shown in Fig. 4.21.

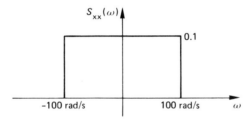

Fig. 4.21

Use Parsevals theorem to determine the mean square value of the time response. Determine the autocorrelation function $R_{xx}(\tau)$ of the signal.

5

Signals in Linear Systems—Time Domain

5.1 Introduction

An important aspect of an engineer's work is a preoccupation with the prediction of the behaviour of some system in response to an excitation or disturbance. For example, an electrical engineer may be interested in the behaviour of a distribution network when a fault appears. A mechanical engineer may be interested in the nature of the vibration set up in a machine tool structure when a cutting operation takes place.

It needs no stretch of the imagination to realise that an excitation or disturbance can be identified as a signal, deterministic or random as the case may be. Also, it is easy to see that the response can be identified as a signal. The engineer wants to predict the character of the response if the character of the excitation and of the system are known. Thus we seek methods whereby this prediction can be effected.

Before embarking on this study we need to clarify certain terms and definitions. For our purposes we will regard a system as some physical object or assembly of physical objects for which we can identify a point at which we can apply an excitation and another point at which we can observe the response. A very simple example would be a mass suspended from a support by a spring. An excitation signal might be a sudden displacement of the support and the response signal might be the consequent displacement of the mass. Another example would be a resistor, connected to a battery via a switch. Closing the switch would apply a step of voltage to the resistor causing a current to flow. The step voltage would be the excitation signal and the current would be the response signal. Many simple and more complicated examples could be cited but this is not necessary here as they will emerge as we progress.

A linear system is one for which the 'size' of the response is proportional to the 'size' of the excitation. For example, the current in the resistor is proportional to the voltage of the battery. This is illustrated graphically in Fig. 5.1.

A consequence of this linear property is the concept of superposition. Consider excitation E_1 giving rise to the response R_1 and E_2 giving rise to R_2. Now take E_3 as the sum of E_1 and E_2, that is $E_3 = E_1 + E_2$. It is easy to perceive that $R_3 = R_1 + R_2$. This means that we can find the response to E_3, that is, E_1 and E_2 added together, by finding the separate responses R_1 and R_2 and adding them together. This is the principle of superposition, a principle that we shall exploit extensively. Clearly it applies to negative values as well as positive values of E and R.

Note that if E and R are not linearly related the principle does not hold, a fact clearly illustrated by Fig. 5.2.

A time-invariant system is one whose characteristics do not change with time. For example, if we excite a system now and get a certain response and then excite it tomorrow in the same way under the same conditions we will get the same response, if it is time-invariant. A time-variant system is one whose characteristics do change with time.

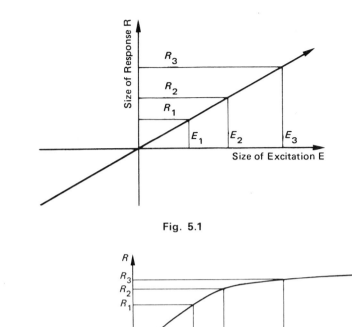

Fig. 5.1

Fig. 5.2

We shall only be concerned with linear-time-invariant (LTI) systems. Further, we shall not be concerned with establishing the relevant characteristics of the systems under study, we shall assume that they are available. The art of finding these characteristics and formulating them mathematically is the subject called mathematical modelling and is not within the scope of this book, however, it will be useful for the reader unfamiliar with modelling to refer to Appendix C.

As always, with any discipline, we meet and use terms which are synonymous or near synonymous and to the uninitiated this can lead to confusion. We draw attention to these as they arise and here are two relevant to the present work.

Excitation signal and input signal are the same and are abbreviated to input for convenience. Response signal and output signal are the same and are abbreviated to output. Also we will consider only realisable systems, these are systems which give an output only when excited. Non-realisable systems include those which can give an output before they are excited but these do not concern us here.

5.2 Input–Output Relationships for Continuous Signals in the Time Domain: Convolution

It is convenient to regard a system as an operator on an input signal which produces an output and our example of a resistor illustrates this. We can regard the resistor as an operator on the voltage to produce the current and Ohm's Law describes the operation,

$$i = \frac{1}{R} \cdot v$$

or $1/R$ operates on v to give i, where R is the resistance, i is the current and v is the potential difference.

Of course, when considering more complicated systems we use the idea of Transfer Function and using the accepted symbols we have

$$Y(s) = G(s) \cdot U(s) \tag{5.1}$$

where $Y(s)$ is the Laplace Transform (LT) of the output y, $G(s)$ is the Transfer Function of the system, (TF), and $U(s)$ is the LT of the input u. This is represented diagrammatically in Fig. 5.3.

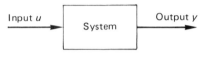

Input u System   Output y

Fig. 5.3

It is important to realise that in using equation (5.1) we have removed the problem from the time domain and put it into the s-domain and for many problems it is far easier to do this than attempt to solve the problem in the time domain, as we shall see. However, this presupposes that the input has a Laplace Transform, or, what amounts to the same thing, that the input is a deterministic signal. However, this is not always the case as, for instance, when the input is a random signal. To handle this situation requires a different method and is one which stays in the time domain (or time world). Recall that when we apply the Laplace Transform to a problem we remove it from the time domain into the s-domain; in other words, time is integrated out by the transformation integral and the complex number s is introduced. When a solution is effected by this method it is often, but not always, necessary to transform it back into the time domain to implement it.

The method which stays in the time world is the convolution method. To study this method we need the idea of the impulse response of a system. This is simply the response of a system to a unit impulse input. To take this further refer to equation (5.1)

$$Y(s) = G(s) \cdot U(s).$$

In the present context u, the input, is a unit impulse, hence its LT, $U(s)$ is unity, that is, $U(s) = 1$. Thus, $Y(s) = G(s) \cdot 1$ or $Y(s) = G(s)$. The output y is then the Inverse LT (ILT) of $Y(s)$ or the ILT of $G(s)$ from which it follows that the impulse response of the system is obtained by taking the ILT of the Transfer Function. We will use the symbol $g(t)$ for impulse response, thus $g(t)$ and $G(s)$ are an LT pair

$$G(s) \leftrightharpoons g(t). \tag{5.2}$$

We are now in a position to develop the convolution method.

To do this we need to refer back to Chapter 1 where we saw that a continuous signal could be made up of a large number of narrow pulses adjacent to one another, see Fig. 1.4. Let the input, u, be so constituted and, for the time being, consider one of the pulses on its own occurring at time $t = \eta_1$. It will have a height $u(\eta_1)$ and duration $\Delta\eta$ and, thus a strength $u(\eta_1)\Delta\eta$ which allows us to use the representation of equation (1.14):

$$u(\eta_1) . \Delta\eta . \delta(t - \eta_1).$$

If this pulse had unit strength and was applied to a system the output would begin at $t = \eta_1$, and be an impulse response $g(t - \eta_1)$ however, its strength is actually $u(\eta_1) . \Delta(\eta)$ and the output will be

$$g(t - \eta_1) . u(\eta_1) . \Delta\eta.$$

These ideas are illustrated in Fig. 5.4.

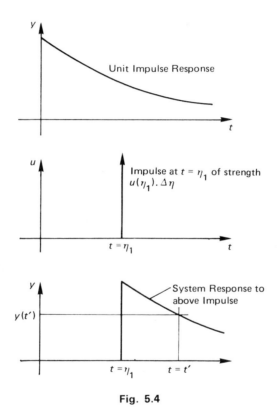

Fig. 5.4

Since we are considering only realisable systems we will not observe any output before $t = \eta_1$ but we will observe output at and after $t = \eta_1$. Thus, at time $t = t'$, $(t' \geqslant \eta_1)$ the output would be $y(t')$.

Now let us consider a second pulse occurring at $t = \eta_2$ as well as that at $t = \eta_1$, $(\eta_2 > \eta_1)$.

It will also produce an impulse response in the same way, proportional to its strength, $u(\eta_2) . \Delta\eta$, which can be represented by

$$g(t - \eta_2) . u(\eta_2) . \Delta\eta.$$

If we again observe y at $t = t'$, $(t' > \eta_2)$ it will have two components, one due to the first impulse and the other due to the second, see Fig. 5.5

$$y(t') = g(t - \eta_1) . u(\eta_1) . \Delta\eta + g(t - \eta_2) . u(\eta_2) . \Delta\eta.$$

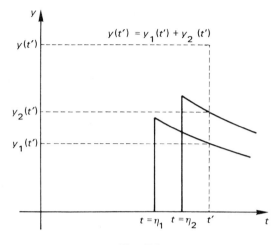

Fig. 5.5

Extending the principle so as to include a large number of pulses we would obtain

$$y(t') = g(t - \eta_1) \cdot u(\eta_1) \cdot \Delta\eta + g(t - \eta_2) \cdot u(\eta_2) \cdot \Delta\eta + \ldots .$$

$$+ \ldots g(t - \eta_{k-1}) \cdot u(\eta_{k-1}) \cdot \Delta\eta + g(t - \eta_k) \cdot u(\eta_k) \cdot \Delta\eta + \ldots .$$

or in summation form:

$$y(t') = \sum_{i=1}^{i=k} g(t - \eta_i) u(\eta_i) \Delta\eta.$$

Now take our input u, made up of a large number of pulses, and let it begin at $t = \eta_0$ and continue indefinitely. If we observe our system output y at $t = t'$ then $y(t')$ will include a contribution from all the impulse responses occurring between $t = \eta_0$ and $t = t'$, that is

$$y(t') = \sum_{i=0}^{i=k} g(t - \eta_i) u(\eta_i) \Delta\eta \tag{5.3}$$

where $\eta_k = t'$.

If we now let the pulses become narrower and narrower so as to approach zero duration then the summation in equation (5.3) becomes an integral:

$$y(t') = \int_{\eta = t_0}^{\eta = t'} g(t - \eta) u(\eta) \mathrm{d}\eta$$

and for any time t this becomes

$$y(t) = \int_{\eta = t_0}^{\eta = t} g(t - \eta) u(\eta) \mathrm{d}\eta \tag{5.4}$$

where $t_0 = \eta_0$.

Equation (5.4) is known as the convolution integral and allows us, in principle, to calculate the response of a system due to a given excitation, the calculation being performed in the time-domain. Its graphical significance is illustrated in Fig. 5.6 from which it is clear that it represents the area under the curve formed from the product of the impulse response $g(t - \eta)$ and the input $u(\eta)$. It is worth studying this diagram carefully in order to grasp the graphical (and hence physical) significance of convolution as well as its algebraic significance as developed above. Note particularly that the first three graphs and the last graph

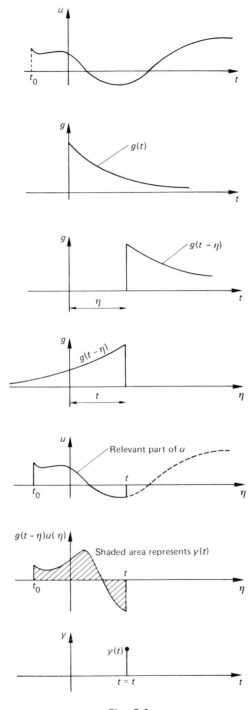

Fig. 5.6

have real time as abscissa but that the others have η as abscissa. This is because the integral gives us y at some particular value of t and therefore t is not a variable as far as the integration is concerned; the integration being with respect to η (the dummy variable). To find y at another value of t requires a re-calculation of the integral with a changed upper limit. Here it is clear that the integration represents the area under the curve formed from the product of $g(t-\eta)$ and $u(\eta)$. Clearly, as t advances the product curve increases in duration but the area under the early parts decreases because the tail of the impulse response decreases. (We are, of course, only considering stable systems whose impulse responses die away with time.) Thus events in the distant past have less weight on the present value of the response. For this reason the impulse response is sometimes known as the weighting function of the system. Fig. 5.7 repeats the latter part of Fig. 5.6 for a later value of t.

Note that the dimension of impulse response is $(\text{Time})^{-1}$ when the dimensions of the input and output are identical. This causes the integrand to have dimensions of (Signal

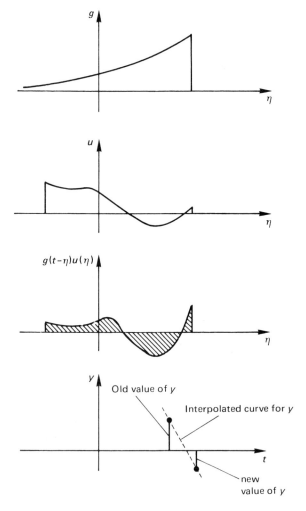

Fig. 5.7

Units) × (Time)$^{-1}$ making the integral, that is the area referred to above, have dimensions of signal units. (If the input and output were not dimensionally the same then the impulse response would include other dimensions to account for this.)

To illustrate the use of the convolution integral let us look at a simple example, one which we will explore from first principles to remind us of the origin of the impulse response and simple linear modelling.

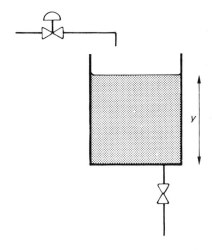

Fig. 5.8

Fig. 5.8 shows a parallel sided, flat based tank which can be filled with water via a control valve. The water can flow out through a valve in the bottom for which we will assume the outflow through it is proportional to the height of water in the tank. (See Appendix C for a justification of this).
Clearly

Input flow rate = output flow rate + Rate of Storage.

Let the input flow rate at any time t be q. Since the outflow is proportional to y we can say that this is Ky, where K is a constant of proportionality, (its dimensions are cubic metres per second per metre of height, i.e. $L^3 T^{-1}/L$ or $L^2 T^{-1}$). The rate of storage is the rate at which the volume of water is increasing which is obviously $A\dfrac{dy}{dt}$, where A is the surface area of the water. Thus we get

$$q = Ky + A\frac{dy}{dt}.$$

If the tank is initially empty, that is $y=0$ when $t=0$ and the control valve opens to admit a steady flow $Q(\text{m}^3 \text{ s}^{-1})$ then we have a first order differential equation to solve. The simplest method of solution is to invoke the method of separation of variables to yield

$$y = \frac{Q}{K}(1 - e^{-t/T})$$

where T is the time constant $\left(T = \dfrac{A}{K}\right)$. y, increasing exponentially.

This basic method of solution is straightforward but does not bring out the ideas we are wanting to illustrate. Nevertheless, before we attack the problem using the convolution integral we will look at the Laplace Transform method for comparison and development of the impulse response.

Taking the LT of the differential equation with the same initial conditions we get

$$Q(s) = K(1 + Ts) Y(s)$$

$$\left(\text{From which the TF is } G(s) = \frac{Y(s)}{Q(s)} = \frac{1}{K(1 + Ts)} \right). \text{ For a step input, } Q, \text{ we get}$$

$$Y(s) = \frac{Q}{Ks(1 + Ts)}$$

and using tables of transforms we would get the same solution as before.

To effect the convolution method we need the impulse response which we found to be the ILT of the TF, $G(s)$. Again, using tables we find

$$\mathscr{L}^{-1} \left[\frac{1}{K(1 + Ts)} \right] \rightarrow \frac{1}{KT} e^{-t/T}$$

$$\text{i.e.} \qquad g(t) = \frac{1}{KT} e^{-t/T}.$$

(Note that the dimensions of $g(t)$ involve T^{-1} but since the output is dimensionally L and the input is $L^3 T^{-1}$ extra dimensions are involved in $g(t)$ yielding dimensions of L^{-2}).

Now let us invoke equation (5.4). Re-writing it using the symbols of the present problem we get

$$y(t) = \int_{t_0}^{t} g(t - \eta) . q(\eta) . d\eta.$$

The input q began at $t = 0$, thus $t_0 = 0$,

$$g(t - \eta) = \frac{1}{KT} e^{-(t - \eta)/T}$$

q is constant at Q therefore

$$y(t) = \int_{0}^{t} \frac{1}{KT} e^{-(t - \eta)/T} Q \, d\eta$$

$$= \frac{Q}{KT} \int_{0}^{t} e^{-(t - \eta)/T} \, d\eta$$

$$= \frac{Q}{KT} \left[\frac{e^{-(t - \eta)/T}}{1/T} \right]_{0}^{t}$$

$$= \frac{Q}{K} \left[e^{-0/T} - e^{-t/T} \right]$$

$$= \frac{Q}{K} [1 - e^{-t/T}].$$

This is, of course, the same result as before. This example is trivial and manageable, however, even with relatively simple deterministic input signals and systems of modest complexity the evaluation of the convolution integral becomes unwieldly. It is for this reason that one always uses an operational method to solve such problems because they are usually far easier. Nevertheless, the concept of convolution is valuable for establishing ideas and is the only way of solving input–output problems for non-deterministic signals when a time solution is required. However, such problems have to be solved computationally because they involve signals which cannot be represented by an explicit function. To evaluate the integral we need an explicit function but when we have such a function it is better to use an operational method! It is emphasised that convolution is essential for developing further ideas in signal processing both for continuous signals and discrete ones.

Equation (5.4) is only one of two forms for the convolution integral. For some problems a second form is more convenient and we will develop this now.

Re-write equation (5.4)

$$y(t) = \int_{\eta=t_0}^{\eta=t} g(t-\eta) . u(\eta) \, d\eta$$

and make the substitution $\lambda = t - \eta$ so that $\eta = t - \lambda$ and $d\eta = -d\lambda$.
Then

$$y(t) = -\int_{\eta=t_0}^{\eta=t} g(\lambda) . u(t-\lambda) \, d\lambda$$

when $\eta = t_0$, $\lambda = t - t_0$ and when $\eta = t$, $\lambda = 0$ and after interchanging the limits of integration to remove the negative sign we get

$$y(t) = \int_{\lambda=0}^{\lambda=t-t_0} g(\lambda) . u(t-\lambda) \, d\lambda$$

when u begins at $t = 0$ both forms of the integral can be simplified:

$$y(t) = \int_0^t g(t-\eta) u(\eta) \, d\eta \tag{5.5}$$

$$y(t) = \int_0^t g(\lambda) . u(t-\lambda) \, d\lambda. \tag{5.6}$$

When studying steady-state problems we let u begin in the infinite past so that the steady state conditions will have been established at time t, for which case $t_0 = -\infty$ and our integrals become respectively

$$y(t) = \int_{-\infty}^t g(t-\eta) u(\eta) \, d\eta \tag{5.7}$$

$$y(t) = \int_0^\infty g(\lambda) u(t-\lambda) \, d\lambda. \tag{5.8}$$

[Using equation (5.7) in our tank problem would give us

$$y(t) = \frac{Q}{KT} \left[\frac{e^{-(t-\eta)/T}}{1/T} \right]_{-\infty}^t$$

$$= \frac{Q}{K} [e^{-0/T} - e^{-(t+\infty)/T}]$$

$$= \frac{Q}{K}$$

which is, of course, the steady state value.]

To illustrate the use of equation (5.6) let us re-work our problem. Using equation (5.6) we get

$$y(t) = \int_0^t \frac{1}{KT} e^{-\lambda/T} Q . H(t-\lambda) \, d\lambda$$

$$= \frac{Q}{KT} \int_0^t H(t-\lambda) e^{-\lambda/T} \, d\lambda.$$

Algebraically this is more difficult to appreciate than before because of the delayed step function $H(t-\lambda)$. Once again we will resort to graphical means to resolve the difficulty,

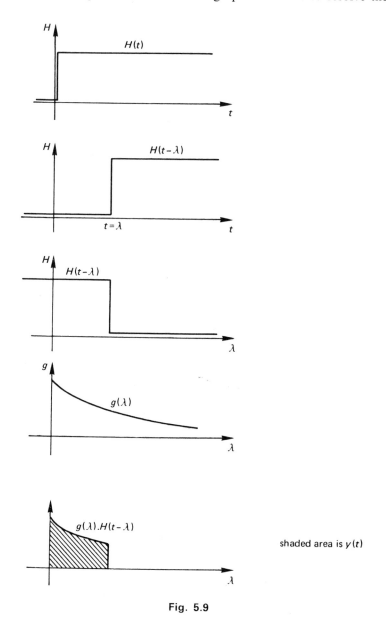

shaded area is $y(t)$

Fig. 5.9

(refer to Fig. 5.9). $H(t)$, the unit step is zero for $t < 0$ and unity for $t > 0$, the delayed step is zero before $t = \lambda$ and unity for $t > \lambda$ as shown. When this unit step is plotted on λ as abscissa it is reversed so that the step is now unity before $\lambda = t$ and zero afterwards. When $H(t - \lambda)$ multiplies $g(\lambda)$, $H(t - \lambda)$ truncates $g(\lambda)$ giving the product curve $[g(\lambda) H(t - \lambda)]$ whose area is $y(t)$.

This area is
$$y(t) = \frac{Q}{KT} \int_0^t e^{-\lambda/T} \, d\lambda$$

$$= \frac{Q}{KT} \left[\frac{e^{-\lambda/T}}{-1/T} \right]_0^t$$

$$= \frac{Q}{K} [e^{0/T} - e^{-t/T}]$$

$$= \frac{Q}{K} [1 - e^{-t/T}].$$

The choice as to whether the form of equation (5.5) or that of (5.6) is best for a particular problem depends on the nature of the problem and experience helps in the decision. It is very rare that one actually uses the convolution integral for the computation of outputs for systems involving continuous signals for the reasons cited earlier but for discrete signals the situation is different, as will be seen.

5.3 Input, Output and Convolution for Discrete Signals

We will now discuss the discrete equivalent of convolution, in other words we will see what modifications must be made to the equations for digital computation.

We begin by recalling equation (5.3), re-written here and re-labelled equation (5.9).

$$y(t') = \sum_{i=0}^{i=k} g(t - \eta_i) u(\eta_i) \Delta\eta \tag{5.9}$$

Whilst this was used in developing the convolution integral it remains an approximation and we will use it as the basis for digital calculation.

Since $\Delta\eta$ is the duration of each constituent pulse of the input signal it makes sense to evaluate y at points in time which are integral multiples of $\Delta\eta$. Doing this would yield the following set of values:

$$y(t_0), \, y(t_0 + \Delta\eta), \, y(t_0 + 2\Delta\eta), \, \ldots \, y(t_0 + i\Delta\eta) \, \ldots \, y(t_0 + K\Delta\eta)$$

which, using the notation of section 1.4, can be written as

$$y[0], \quad y[1], \quad y[2], \ldots y[i], \ldots y[k]$$

$$\text{or} \quad y_0, \quad y_1, \quad y_2 \ldots y_i, \ldots y_k.$$

In the same way g and u can be written, respectively, as

$$g[k], \quad g[k-1], \quad g[k-2], \ldots g[k-i], \ldots g[0]$$

$$\text{or} \quad g_k, \quad g_{k-1}, \quad g_{k-2}, \ldots g_{k-i}, \ldots g_0$$

$$u[0], \quad u[1], \quad u[2], \ldots u[i], \ldots u[k]$$

$$\text{or} \quad u_0, \quad u_1, \quad u_2, \ldots u_i, \ldots u_k.$$

The choice between the use of the bracket notation, $x[i]$, or the suffix notation x_i, is arbitrary but by and large the former will be used for developing principles via equations and the latter in routine computational working.

A further point on notation. Since $\Delta\eta$ is a common multiplier in the convolution sum equation (5.9) we will combine it with u so that henceforth, merely for convenience, we will understand that $u[i]$ means $u(\eta_i) \cdot \Delta\eta$. If we forget this, our calculations will be wrong when they are compared with those that might have been obtained using the convolution integral and continuous signals.

We can now write equation (5.9) as

$$y[k] = \sum_{i=0}^{k} g[k-i]\, u[i] \tag{5.10a}$$

$$\text{or} \qquad y_k = \sum_{i=0}^{k} g_{k-i}\, u_i. \tag{5.10b}$$

Obviously we can put these in their alternative forms

$$y[k] = \sum_{i=0}^{k} g[i]\, u[k-i] \tag{5.11a}$$

$$\text{or} \qquad y_k = \sum_{i=0}^{k} g_i\, u_{k-i}. \tag{5.11b}$$

We will call $g[i]$ the discrete unit impulse response or the unit pulse response for short.

We can represent this graphically as shown in Fig. 5.10 where each signal is now regarded as a train of pulses, but bearing in mind the way in which we have derived our convolution sum [equations (5.9)–(5.11)] through the use of the unit impulse function, $\delta(t)$, we must remember that the height of each 'pulse', u_i involves strength and not just amplitude. To aid understanding we will imagine that the continuous system and its continuous signals are replaced by a discrete system and discrete signals as illustrated in Fig. 5.11. It is therefore reasonable to regard $g[i]$ as the response of a digital system to a single unit pulse input $\delta[i]$.

To illustrate the use of equations (5.10) and (5.11) we will work our tank problem, but to make the calculations easier we will give numerical values to the parameters. Let $Q = 10 \text{ m}^3 \text{ s}^{-1}$, $K = 0.5 \text{ m}^2 \text{ s}^{-1}$, $T = 1 \text{ s}$ and we will make $\Delta\eta$ small in comparison to T at $\Delta\eta = 0.1 \text{ s}$.

Using these values

$$g(t) = \frac{1}{KT}\, \mathrm{e}^{-t/T} = 2\mathrm{e}^{-t} \qquad \text{and} \qquad t = \Delta\eta \cdot i = 0.1i$$

from which,

$$g[i] = 2\mathrm{e}^{-0.1i} \qquad i = 1, 2, 3, \ldots$$

and

$$g_0 = 2\mathrm{e}^{-0} = 2$$

$$g_1 = 2\mathrm{e}^{-0.1} = 1.8097$$

$$g_2 = 2\mathrm{e}^{-0.2} = 1.6375$$

$$g_3 = 1.4816$$

$$g_4 = 1.3406.$$

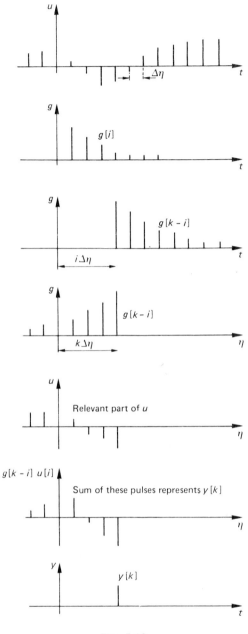

Fig. 5.10

The input is at constant amplitude $10 \, \mathrm{m^3 \, s^{-1}}$ and we get, remembering to include $\Delta \eta$,

$$u_0 = 10 \times 0.1 = 1$$

$$u_1 = 1.$$

$$u_2 = 1.\text{etc.}$$

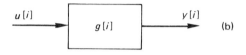

Fig. 5.11

Now using equation (5.10b) we get

$$y_0 = \sum_{i=0}^{0} g_{0-i} u_i = g_{0-0} \cdot u_0$$

$$= 2 \times 1 = 2$$

$$y_1 = \sum_{i=0}^{1} g_{1-i} u_i = g_{1-0} u_0 + g_{1-1} \cdot u_1$$

$$= g_1 u_0 + g_0 u_1$$

$$= 1.8097 \times 1 + 2 \times 1 = 3.8097$$

$$y_2 = \sum_{i=0}^{2} g_{2-i} u_i = g_{2-0} u_0 + g_{2-1} u_1 + g_{2-2} u_2$$

$$= g_2 u_0 + g_1 u_1 + g_0 u_2$$

$$= 1.6375 \times 1 + 1.8097 \times 1 + 2 \times 1 = 5.4472$$

$$y_3 = g_3 u_0 + g_2 u_1 + g_1 u_2 + g_0 u_3$$

$$= 1.4816 \times 1 + 1.6375 \times 1 + 1.8097 \times 1 + 2 \times 1$$

$$= 6.9288$$

using equation (5.11b), we would have obtained

$$y_0 = g_0 u_0$$

$$= 2 \times 1 = 2$$

$$y_1 = g_0 u_1 + g_1 u_0$$

$$= 3.8097$$

$$y_2 = g_0 u_2 + g_1 u_1 + g_2 u_0$$

$$= 5.4472.$$

Clearly the two methods give the same results (as they should!) and if the calculations were continued indefinitely we would get the complete discrete response.

It is interesting to compare these values with those obtained for the original continuous system for which

$$y(t) = \frac{Q}{K}(1 - e^{-t/T})$$

$$= 20(1 - e^{-t}).$$

Continuous	**Discrete**
$y(0) = 0$	$y_0 = 2$
$y(0.1) = 1.903$	$y_1 = 3.81$
$y(0.2) = 3.625$	$y_2 = 5.447$
$y(0.3) = 5.184$	$y_3 = 6.929$
$y(0.4) = 6.594$	

The discrepancy arises from the fact that the input, u, for the discrete system is a train of weighted impulses and when the first one, u_0, arrives it excites the system which produces its impulse response which, for this first order system, has a non-zero initial value. During the interval between u_0 and u_1 the response decays but on the arrival of u_1 the system is again excited and so on, giving an actual response as shown in Fig. 5.12. Discrete convolution only gives us the response values at the instant of each input pulse.

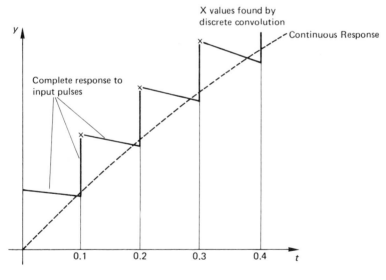

Fig. 5.12

An alternative approximation to the continuous solution can be obtained by the following device. When a discontinuous signal, such as a step, has its discontinuity at a sampling point there is ambiguity as to its value at that instant; does it take the value before the discontinuity or does it take the value afterwards? (For a unit step do we take zero or unity?) Compromising we take the halfway value, for example g would become

$$g_0 = 1$$
$$g_1 = 1.8097$$
$$g_2 = 1.6375$$

and
$$u_0 = 0.5$$
$$u_1 = 1$$
$$u_2 = 1.$$

(A more rigorous justification for the halfway choice is possible but it is beyond the scope of this book).

The train of impulses exciting the linear system can be embraced by the arrangement shown in Fig. 5.13, in which the switch (known as the sampler) closes momentarily and regularly at an interval of time $\Delta\eta$ to cause the continuous input u to be converted into a train of weighted impulses $u^*(t)$ which then form the input to the system.

Fig. 5.13

From the worked examples it is clear that there is a pattern to the sums of products and these are shown in the following equations

$$y_0 = g_0 u_0$$
$$y_1 = g_1 u_0 + g_0 u_1$$
$$y_2 = g_2 u_0 + g_1 u_1 + g_0 u_2$$
$$y_3 = g_3 u_0 + g_2 u_1 + g_1 u_2 + g_0 u_3.$$

It is perfectly reasonable to invent values of g for negative argument, g_{-1}, g_{-2}, g_{-3} etc. and for our type of system they are all zero so inclusion with those above will make no difference:

$$y_0 = g_0 u_0 + g_{-1} u_1 + g_{-2} u_2 + g_{-3} u_3 + \ldots$$
$$y_1 = g_1 u_0 + g_0 u_1 + g_{-1} u_2 + g_{-2} u_3 + \ldots$$
$$y_2 = g_2 u_0 + g_1 u_1 + g_0 u_2 + g_{-1} u_3 + \ldots$$
$$y_3 = g_3 u_0 + g_2 u_1 + g_1 u_2 + g_0 u_3 + \ldots$$

If we now define $[y]$ as the column vector of y values we have

$$[y] = \begin{bmatrix} y_0 \\ y_1 \\ \vdots \end{bmatrix}$$

Doing the same for u gives

$$[y] = [G][u] \tag{5.12}$$

where $[G]$ is the matrix
$$\begin{bmatrix} g_0 & g_{-1} & g_{-2} \cdots \\ g_1 & g_0 & \cdots \\ g_2 & g_1 & \cdots \end{bmatrix}$$

Computationally we have expressed discrete convolution as a matrix operation which could be effected readily by a computer. However, it is a slow operation especially as all the data must be read in before we get any values of y calculated.

Obviously the alternative form of discrete convolution will give a similar matrix operation and will appear as

$$[y] = [U][g] \tag{5.13}$$

where y is as before, $[g] = \begin{bmatrix} g_0 \\ g_1 \\ \vdots \\ g_n \end{bmatrix}$

and $[U] = \begin{bmatrix} u_0 & u_{-1} & u_{-2} \dots \\ u_1 & u_0 & u_{-1} \dots \\ u_2 & u_1 & u_0 \dots \\ \dots \end{bmatrix}$

in which u_{-1}, u_{-2}, u_{-3}, etc. are all zero in this case.

5.4 Input, Output and Difference Equations

So far our work on linear systems has concentrated on the idea that the system is an operator on an input to produce an output. We shall continue to follow this idea later but before we do so it is important and appropriate now to look at a slightly different approach. We have, in passing, already met the principle when we used a differential equation, in the solution of our tank problem. See page 107. There is no reason why we should not set up the relevant differential equations for more complicated LTI systems and attempt to find their outputs by invoking classical methods, the D-operator method or the LT method. We will not discuss these methods since they have been written about extensively, however, it is important to remember their existance and their place in the hierarchy of solution methods, bearing in mind that they are relevant to continuous systems. What does concern us here is their counterpart for discrete signals.

The equivalent of a differential equation for discrete systems is the difference equation and it is this that we discuss briefly here. There is a whole literature relevant to them as there is for differential equations; all we require is sufficient skill for handling, solving and as a basis for further reading. Whilst they can be regarded as a subject in their own right we are concerned with their use as a representation of the input–system–output trio and, because of this, we will show how to obtain a difference equation representation from the differential equation representation.

Consider the arrangement shown in Fig. 5.11(a) and let the system be represented by the first order differential equation

$$a_1 \frac{dy}{dt} + a_0 y = b_0 u.$$

If we now sample the signals y and u, as in Fig. 1.2, we would get the sequences

$$y(t_0),\ y(t_1) \dots y(t_i)$$

$$u(t_0),\ u(t_1) \dots u(t_i)$$

or in simpler notation

$$y_0, y_1, \ldots y_i$$

$$u_0, u_1, \ldots u_i.$$

Clearly we could approximate $\dfrac{dy}{dt}$ by $\dfrac{y_{i+1} - y_i}{\Delta}$ and re-write our differential equation in an approximate form:

$$a_1 \left(\frac{y_{i+1} - y_i}{\Delta} \right) + a_0 y_i = b_0 u_i$$

in which Δ is the time between samples. Re-arranging this we get

$$a_1 y_{i+1} + (a_0 \Delta - a_1) y_i = b_0 \Delta u_i$$

or

$$\alpha_1 y_{i+1} + \alpha_0 y_i = \beta_0 u_1 \tag{5.14}$$

in which $\alpha_1 = a_1$, $\alpha_0 = a_0 \Delta - a_1$ and $\beta_0 = b_0 \Delta$.

Here we use 'forward differences'—an alternative approach is to use 'backward differences' as follows:

Let

$$\frac{dy}{dt} \simeq \frac{y_i - y_{i-1}}{\Delta}$$

then the differential equation becomes

$$a_1 \left(\frac{y_i - y_{i-1}}{\Delta} \right) + a_0 y_i = b_0 \Delta u_i$$

leading to

$$(a_1 + \Delta a_0) y_i - a_1 y_{i-1} = \Delta b_0 u_i$$

and slightly different results.

Equation (5.14) is a difference equation. It is possible to rearrange it and obtain a form which will allow us to obtain an alternative solution. We can get

$$y_{i+1} = -\frac{\alpha_0}{\alpha_1} y_i + \frac{\beta_0}{\alpha_1} u_i.$$

If we know the parameters α_1, α_0, and β_0 and the present values y_i and u_i we can calculate the next value of y namely y_{i+1}. Having found this we can call it the present value and using the equation with a new present value of u we can calculate the next value of y. Repeating this process we produce the sequence of values for y from the system parameters and the sequence of values for u. Let us illustrate this using the example of the tank.

We have

$$A \frac{dy}{dt} + Ky = q$$

in which $\dfrac{A}{K} = T = 1\,\text{s}$, $K = 0.5\,\text{m}^2\,\text{s}^{-1}$ and q is constant at $10\,\text{m}^3\,\text{s}^{-1}$. We find A to be $0.5\,\text{m}^2$ and the differential equation to be

$$0.5 \frac{dy}{dt} + 0.5y = 10.$$

Taking Δ to be $0.1\,\text{s}$ gives the difference equation

$$0.5 y_{i+1} - 0.45 y_i = 0.1 u_i.$$

Re-arranging to the iterative form gives

$$y_{i+1} = 0.9y_i + 0.2u_i.$$

Taking the same initial conditions as before we get $y_i = 0$ when $i = 0$, that is $y_0 = 0$. The sequence u is $u_0 = 10$, $u_1 = 10$, $u_2 = 10$, etc.
Then

$$y_{0+1} = y_1 = 0.9y_0 + 0.2u_0$$

$$= 0.9 \times 0 + 0.2 \times 10 = 2$$

$$y_{1+1} = y_2 = 0.9y_1 + 0.2u_1$$

$$= 0.9 \times 2 + 0.2 \times 10 = 3.8$$

$$y_{2+1} = y_3 = 0.9 \times 3.8 + 0.2 \times 10 = 5.42$$

$$y_4 = 6.87.$$

These, of course, compare with previous results and tally reasonably with those of the solution from the differential equation, however, this is because we have chosen $y_0 = 0$ as an initial condition. Had we chosen $y_0 = 2$ the results would have tallied closely with those from the discrete convolution method.

We can apply the same principles to higher order equations and to illustrate the method we will look at the second order equation:

$$a_2 \frac{d^2y}{dt^2} + a_1 \frac{dy}{dt} + a_0 y = b_1 \frac{du}{dt} + b_0 u.$$

In anticipation of later work and for completeness we have included the first derivative of u. Remembering that the second derivative is the rate of change of a rate of change we can approximate as follows

$$\frac{d^2y}{dt^2} \simeq \frac{\dfrac{dy_{i+1}}{dt} - \dfrac{dy_i}{dt}}{\Delta}$$

$$\simeq \frac{\left[\dfrac{y_{i+2} - y_{i+1}}{\Delta}\right] - \left[\dfrac{y_{i+1} - y_i}{\Delta}\right]}{\Delta}$$

$$= \frac{y_{i+2} - 2y_{i+1} + y_i}{\Delta^2}.$$

Substituting this into the differential equation together with the approximations for $\dfrac{dy}{dt}$ and $\dfrac{du}{dt}$ we get*

$$a_2 y_{i+2} + (a_1 \Delta - 2a_2)y_{i+1} + (a_0 \Delta^2 - a_1 \Delta + a_2)y_i = \Delta b_1 u_{i+1} + (\Delta^2 b_0 - \Delta b_1)u_i.$$

In the same way as before we can re-define the coefficients and obtain

$$\alpha_2 y_{i+2} + \alpha_1 y_{i+1} + \alpha_0 y_i = \beta_1 u_{i+1} + \beta_0 u_i. \qquad (5.15)$$

* Again we use forward differences, but backward difference can also be used.

Once again this can be re-arranged to an iterative form and a sequence for y obtained, but this time two initial conditions will be required, namely y_0 and y_1:

$$y_{i+2} = -\frac{\alpha_1}{\alpha_2}y_{i+1} - \frac{\alpha_0}{\alpha_2}y_i + \frac{\beta_1}{\alpha_2}u_{i+1} + \frac{\beta_0}{\alpha_2}u_i.$$

Clearly, we can apply the method and convert the nth order differential equation

$$a_n\frac{d^n y}{dt^n} + a_{n-1}\frac{d^{n-1}y}{dt^{n-1}} + \ldots\ldots a_0 y = b_m\frac{d^m u}{dt^m} + \ldots b_0 u,$$

into an nth order difference equation:

$$\alpha_n y_{i+n} + \alpha_{n-1}y_{i+n-1} + \ldots \alpha_0 y_i = \beta_n u_{i+m} + \ldots \beta_0 u_i. \tag{5.16}$$

Just as it is necessary to have n initial conditions for an nth order differential equation we need n initial conditions for an nth order difference equation. Also just as there are operational methods for solving differential equations there are operational methods for solving difference equations. We will examine this point later.

It is emphasised that difference equations (sometimes known as recurrence relations) can stand in their own right without any connection with differential equations but our reason for deriving them in this way is to show how they can represent physical systems, which is our concern. Further, it must be appreciated that the discrete approximations to derivatives are rather crude and that there are more accurate ways of solving differential equations numerically which can be found in any book on Numerical Methods.

To round off this section let us work an iterative solution to a second order equation:

$$4\frac{d^2 y}{dt^2} + 5\frac{dy}{dt} + y = 10u.$$

Using differences we get

$$\frac{4}{\Delta}\left(\frac{y_{i+2} - y_{i+1}}{\Delta} - \frac{y_{i+1} - y_i}{\Delta}\right) + 5\left(\frac{y_{i+1} - y_i}{\Delta}\right) + y_i = 10u_i.$$

Once again, take u to be a step of unit height yielding the train of pulses

$$u_0 = 1,\ u_1 = 1,\ u_2 = 1\ \text{etc.}$$

Using this and simplifying gives the difference equation:

$$4y_{i+2} - 7.5y_{i+1} + 3.51y_i = 0.1 \qquad \text{for } \Delta = 0.1$$

and $$y_{i+2} = 1.875y_{i+1} - 0.8775y_i + 0.025.$$

Taking zero initial conditions, that is, $y_0 = 0$ and $y_1 = 0$ produces

$y_0 = 0$
$y_1 = 0$
$y_2 = 1.1875 \times 0 - 0.8775 \times 0 + 0.025 = 0.025$
$y_3 = 1.875 \times 0.025 - 0.8775 \times 0 + 0.025 = 0.071\,875$
$y_4 = 1.875 \times 0.071\,88 - 0.8775 \times 0.025 + 0.025 = 0.137\,828$
$y_5 = 0.220\,357$
$y_6 = 0.317\,226.$

As an exercise find the exact values by solving the differential equation and evaluating at intervals of 0.1 seconds.

5.5   The z-Transform

For continuous systems input–output problems have been solved by the use of the Laplace Transform as has the solution of constant coefficient linear differential equations. Recall that this Transform is an integral operation which integrates out time t, and introduces the complex variable, s, in other words, the problem in hand is transformed from the t-domain into the s-domain. From a practical point of view this is effected to make problems easier. Thus it is not unreasonable to suppose that we can do the same for discrete problems. Indeed a relevant transformation is possible and has many properties which correspond to properties of the (Laplace Transform) LT, it is the z-Transform (ZT). The LT of the function $f(t)$ is defined as

$$F(s) = \int_0^\infty f(t) e^{-st} dt$$

and in a similar way we could boldly define the z-transform of the sequence $[x_i]$ as

$$X(z) = \sum_{i=0}^{i=\infty} x_i z^{-i}.$$

However, its relationship with the LT is so simple and elegant that we will show how it is derived therefrom.

The LT of the unit impulse function $\delta(t)$ is unity, and that of a unit impulse delayed by an amount Δ is $e^{-s\Delta}$, see Table 5.1.

Table 5.1

Delayed Impulse	LT
$\delta(t)$	1
$\delta(t-\Delta)$	$e^{-s\Delta}$
$\delta(t-2\Delta)$	$e^{-s2\Delta}$
$\vdots$	
$\delta(t-i\Delta)$	$e^{-si\Delta}$

Recalling equation (1.16) we can write

$$x^*(t) = x(0)\Delta\eta\delta(t) + x(\Delta)\Delta\eta\delta(t-\Delta) + x(2\Delta)\Delta\eta\delta(t-2\Delta) + \ldots$$
$$\ldots x(i\Delta)\Delta\eta\,\delta(t-i\Delta)\ldots.$$

Remember that Δ is a specific time delay whereas $\Delta\eta$ is the duration of the pulse.

Now take the LT of $x^*(t)$ and obtain

$$X^*(s) = x(0)\Delta\eta + x(\Delta)\Delta\eta e^{-s\Delta} + x(2\Delta)\Delta\eta e^{-2s\Delta} + \ldots x(i\Delta)\Delta\eta e^{-si\Delta} \ldots$$
$$= \Delta\eta \sum_{i=0}^{\infty} x(i\Delta) e^{-si\Delta}.$$

Since $\Delta\eta$ is a multiplier it can be subsumed into the value of x and using discrete notation we get

$$X^*(s) = \sum_{i=0}^{\infty} x_i e^{-si\Delta}.$$

Now define $z = e^{s\Delta}$ to get

$$X(z) = \sum_{i=0}^{\infty} x_i z^{-i}$$

in which we have written $X(z)$ instead of $X^*(s)$ (since s does not appear on the r.h.s.) and forms a defining equation for the z-transform. What does it mean? It means that if we have a sequence $[x_i]$ we can transform it into a function and perform manipulations for problem solving including, of course, input–output problems. Obviously an example is called for, let us find the ZT of the step sequence $H[i]$

$$H(i) = 1 \qquad \text{for } i \geq 0$$

$$= 0 \text{ for } \qquad i < 0$$

$$X(z) = \sum_{i=0}^{\infty} 1 \cdot z^{-i}$$

$$= 1 + z^{-1} + z^{-2} + \ldots$$

This series is that obtained by performing the division $\dfrac{z}{z-1}$. Thus

$$X(z) = \frac{z}{z-1} \qquad \text{for } H[i].$$

We could now fill several square metres of paper working out the ZT of lots of sequences but since this has already been done and produced as tables and also because we are more concerned with using the transform as a tool it is not necessary. The ZT is discussed in greater detail in Appendix B, where further examples are worked. For the time being we will record a few simple transform pairs in Table 5.2 for use in the examples to be worked later. A longer table will be found in Table A2, p 303.

Note that if we put $e^{-a\Delta} = \phi$, we simplify to $e^{-ai\Delta} = \phi^i$ and third pair can be written $\phi^i \rightleftharpoons \dfrac{z}{z-\phi}$ and the fourth pair as $1 - \phi^i \rightleftharpoons \dfrac{(1-\phi)z}{(z-1)(z-\phi)}$. Before we go on to use the ZT as an operational method we will quote one or two useful properties that will be required for the later development. Their proof and other properties are given in Appendix B.

The ZT of the sum (or difference) of a number of sequences

$$\mathscr{Z}\{a[w_i] \pm b[x_i] \pm c[y_i] \pm \ldots\}$$
$$= aW(z) \pm bX(z) \pm cY(z) \pm \ldots$$

i.e. the transform of a sum is the sum of the individual transforms a, b and c are constants.

The ZT of a shifted sequence (forward shift)

$$\mathscr{Z}\{[x_{i+j}]\} = z^j X(z) - (z^j x_0 + z^{j-1} x_1 + \ldots z x_{j-1})$$

in which j is the number of intervals of shift. Property 1 has its direct counterpart in the world of LT. Property 2 is the counterpart of the differentiation property of LT:

$$\mathscr{L}\left(\frac{d^n x}{dt^n}\right) = s^n X(s) - \left[s^{n-1}x(0) + s^{n-2}\frac{dx(0)}{dt} + \ldots \frac{d^{n-1}x(0)}{dt^{n-1}}\right].$$

Again, as with the LT, we shall find it necessary to perform inverse transformations that is, given a ZT find the inverse z-transform to obtain the corresponding sequence. We will use

Table 5.2

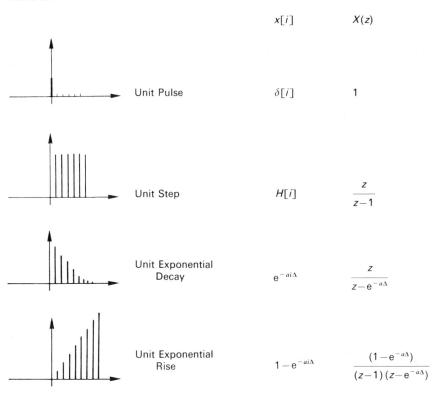

	x[i]	X(z)
Unit Pulse	$\delta[i]$	1
Unit Step	$H[i]$	$\dfrac{z}{z-1}$
Unit Exponential Decay	$e^{-ai\Delta}$	$\dfrac{z}{z-e^{-a\Delta}}$
Unit Exponential Rise	$1-e^{-ai\Delta}$	$\dfrac{(1-e^{-a\Delta})}{(z-1)(z-e^{-a\Delta})}$

the idea of partial fractions, (but with a slight difference) and the table of transforms. We expand $X(z)$ as

$$X(z) = A_1 \frac{z}{z-\phi_1} + A_2 \frac{z}{z-\phi_2} + \ldots A_n \frac{z}{z-\phi_n}$$

Then

$$x = A_1 \phi_1^i + A_2 \phi_2^i + \ldots A_n \phi_n^i.$$

As usual we will find that the a_n can be real or complex and that repeated roots have to be handled but these points will be apparent in the examples.

5.6   The Pulse Transfer Function

We were reminded of the idea of transfer function in section 5.2 and equation (5.1) $[Y(s) = G(s) . U(s)]$ gives the relationship between the input $U(s)$, the output, $Y(s)$ and the TF, $G(s)$. Using the ZT we can get a similar relationship for systems with a train of pulses as input. Let the input to a continuous system, whose TF is $G(s)$, be

$$u^*(t) = u_0\delta(t) + u_1\delta(t-\Delta) + u_2\delta(t-2\Delta) + \ldots$$

then

$$U^*(s) = u_0 + u_1 e^{-s\Delta} + u_2 e^{-s2\Delta} + \dots$$

in which, as usual, u_i is the strength of the respective pulse. Then

$$Y(s) = G(s) . U^*(s)$$
$$= G(s)u_0 + G(s)u_1 e^{-s\Delta} + G(s)u_2 e^{-s2\Delta} + \dots$$

and

$$y(t) = g(t)u_0 + g(t-\Delta)u_1|_{t \geqslant \Delta} + g(t-2\Delta)u_2|_{t \geqslant 2\Delta} + \dots$$
$$y(i\Delta) = g(i\Delta)u_0 + g(i\Delta - \Delta)u_1|_{i \geqslant 1} + g(i\Delta - 2\Delta)u_2|_{i \geqslant 2} + \dots$$

Taking ZT we get

$$Y(z) = G(z)u_0 + G(z)u_1 z^{-1} + G(z)u_2 z^{-2} + \dots$$
$$= G(z)(u_0 + u_1 z^{-1} + u_2 z^{-2} + \dots$$
$$= G(z) . U(z)$$

$G(z)$ is the pulse transfer function (PTF).

Taking our previous example (section 5.3) to illustrate we have $g(t) = 2e^{-t}$ or $g[i] = e^{-0.1i}$ which gives $G(z) = \dfrac{2z}{z - e^{-0.1}} = \dfrac{2z}{z - 0.9048}$. Then the output is $Y(z) = \dfrac{2z}{z - 0.9048} U(z)$ and since the input is a step of height 10 units we get

$$U(z) = \frac{10z}{z-1} . \Delta = \frac{10z}{z-1} \times 0.1$$

giving

$$Y(z) = \frac{2z^2}{(z-1)(z-0.9048)}.$$

Using Table 5.2 we get

$$y[i] = \frac{2}{1 - 0.9048}(1 - 0.9048^{i+1})$$
$$= 21.08(1 - 0.9048^{i+1})$$

Note that

$$\frac{2z}{(z-1)(z-0.9048)} \to 21.08(1 - 0.9048^i)$$

but $Y(z)$ has z^2 in the numerator rather than z which means that we have to advance the latter response by one sampling interval.

5.7 Convolution and Laplace Transformation

It should now be readily appreciated that convolution in the time domain is the counterpart of multiplication in the s-domain since we obtain an output in the former by

$$y(t) = \int_0^t u(\eta) . g(t-\eta) . d\eta$$

and in the latter by

$$Y(s) = G(s) . U(s).$$

We will now prove that the LT of convolution is multiplication

$$\mathscr{L}\left[y(t)\right]=\mathscr{L}\left[\int_0^t u(\eta).g(t-\eta).d\eta\right]$$

$$=\mathscr{L}\left[\int_0^\infty u(\eta).H(t-\eta).g(t-\eta)d\eta\right]$$

$$=\int_0^\infty\left[\int_0^\infty H(t-\eta).g(t-\eta).u(\eta).d\eta\right]e^{-st}dt$$

$$=\int_0^\infty\int_0^\infty H(t-\eta).g(t-\eta)u(\eta)e^{-st}e^{-s\eta}e^{s\eta}d\eta\,dt$$

$$=\int_0^\infty\int_0^\infty H(t-\eta)g(t-\eta)e^{-s(t-\eta)}u(\eta).e^{-s\eta}d\eta\,dt.$$

Now make the substitution $\lambda=t-\eta$ and obtain

$$\mathscr{L}\left[y(t)\right]=\int_{-\eta}^\infty\int_0^\infty H(\lambda).g(\lambda)e^{-s\lambda}u(\eta)e^{-s\eta}d\eta\,d\lambda$$

$$=\int_0^\infty\int_0^\infty g(\lambda)e^{-s\lambda}u(\eta).e^{-s\eta}d\eta\,d\lambda.$$

This double integral can now be expressed as the product of two integrals because of the independence of the variables λ and η

$$\mathscr{L}\left[y(t)\right]=\int_0^\infty g(\lambda).e^{-s\lambda}d\lambda.\int_0^\infty u(\eta).e^{-s\eta}d\eta$$

in which it will be noticed that each integral defines a LT, therefore

$$\mathscr{L}\left[y(t)\right]=Y(s)=G(s).U(s). \tag{5.17}$$

This relationship forcefully shows why we always solve deterministic problems via transformation—it is easier to multiply than to integrate! Note how the step function $H(\ )$ has been used to simplify the manipulation of the limits of integration.

5.8   Convolution and the z-Transformation

We will now show the same relationship for discrete signals

$$\mathscr{Z}\left[y[i]\right]=\mathscr{Z}\left[\sum_{i=0}^j q[j-i]u[i]\right]$$

$$=\mathscr{Z}\left[\sum_{i=0}^\infty H[j-i]g[j-i]u[i]\right]$$

$$=\sum_{j=0}^\infty\left[\sum_{i=0}^\infty H[j-i]g[j-i]u[i]\right]z^{-j}.$$

Now make the substitution $k=j-i$ and obtain, in a similar way to that in section 5.7,

$$\mathscr{Z}[y[i]] = \sum_{k=-i}^{\infty} \sum_{i=0}^{\infty} H[k]g[k]u[i]z^{-(k+i)}$$

$$= \sum_{k=0}^{\infty} \sum_{i=0}^{\infty} g[k]u[i]z^{-k}z^{-i}.$$

For the same reason as before the double sum can be expressed as a product of sums:

$$\mathscr{Z}[y[i]] = \sum_{0}^{\infty} g(k) \cdot z^{-k} \sum_{0}^{\infty} u[i]z^{-i}$$

or
$$Y(z) = G(z) \cdot U(z). \tag{5.18}$$

Again note the use of the step function $H[\ \ ]$.

5.9 Illustrative Example

Except in one case our worked examples have been relevant to a first order system so in order to extend the ideas to higher order systems and pull together the above principles we will now work an example from a second order system. Our earlier example of a second order equation was

$$4\frac{d^2y}{dt^2} + 5\frac{dy}{dt} + y = 10u.$$

(a) Let us assume this is the model of a system for which y is the output and u is the input. Being a continuous system we can invoke LT methods and obtain, for zero initial conditions,

$$(4s^2 + s + 1)Y(s) = 10u(s)$$

from which
$$Y(s) = \frac{10}{4s^2 + 5s + 1}U(s)$$

when $U(s)$ is the unit step input

$$Y(s) = \frac{10}{4s^2 + 5s + 1} \cdot \frac{1}{s}$$

$$= \frac{10}{s(4+\mathfrak{J})(s+1)}$$

$$= \frac{2.5}{s(s+0.25)(s+1)}.$$

Note that the two time constants are $4s$ and $1s$. Using a table of transforms (and partial fractions if necessary) we can get:

$$y(t) = 10(1 - \tfrac{4}{3}e^{-t/4} + \tfrac{1}{3}e^{-t}).$$

The TF is, of course, $G(s) = \dfrac{10}{4s^2 + 5s + 1} = \dfrac{2.5}{(s+0.25)(s+1)}$ from which the impulse response is

$g(t) = 10/3(e^{-t/4} - e^{-t})$ and using the convolution integral we get

$$y(t) = \int_0^t \frac{10}{3} (e^{-(1/4)(t-n)} - e^{-(t-n)}). 1. dn$$

$$= \frac{10}{3} \left[\frac{e^{-(1/4)(t-n)}}{1/4} - e^{-(t-n)} \right]_0^t$$

$$= \tfrac{10}{3} [4 - 1 - 4e^{-t/4} + e^{-t}]$$

$$= 10[1 - \tfrac{4}{3}e^{-t/4} + \tfrac{1}{3}e^{-t}] \text{ as before.}$$

(b) We will now look at the problem using discrete methods, taking firstly the ZT method. Since the impulse response is

$$g(t) = \tfrac{10}{3}(e^{-t/4} - e^{-t})$$

we can easily obtain the pulse response $g[i]$ if we specify a suitable sampling interval (that is the interval between pulses). We will take this to be less than or equal to one tenth of the shortest time constant, let us take $\Delta = 0.1$ s.
Then

$$g[i] = \tfrac{10}{3}(e^{-0.1i/4} - e^{-0.1i})$$

$$= \tfrac{10}{3}(0.9753^i - 0.9048^i).$$

The PTF is then

$$G(z) = \frac{10}{3} \left(\frac{z}{z - 0.9753} - \frac{z}{z - 0.9048} \right)$$

$$U(z) = \frac{z}{z-1} \times 0.1 \qquad \text{and} \qquad Y(z) = G(z). U(z)$$

i.e.

$$Y(z) = \frac{1}{3} \left(\frac{z^2}{(z-1)(z-0.9753)} - \frac{z^2}{(z-1)(z-0.9048)} \right).$$

We will invoke partial fractions on each of the two principal terms but since they are both of the same form the same method will do for each.

Let

$$\frac{z^2}{(z-1)(z-a)} = \frac{Az}{z-1} + \frac{Bz}{z-a}$$

or

$$z^2 = Az(z-a) + Bz(z-1).$$

Put $z = 1$ Then

$$1 = A(1-a) \qquad \text{giving } A = \frac{1}{1-a}.$$

Put $z = a$ Then

$$a^2 = Ba(a-1) \qquad \text{giving } B = \frac{a}{a-1}.$$

For the first term in brackets for $Y(z)$ we get,

$$A = \frac{1}{1-0.9753} = 40.4858$$

$$B = \frac{0.9753}{0.9753-1} = -39.4858.$$

For the second term we get

$$A = 10.5042$$

$$B = -9.5042.$$

Then

$$Y(z) = \frac{1}{3}\left[\frac{40.4858z}{z-1} - \frac{39.4858z}{z-0.9753} - \frac{10.5042z}{z-1} + \frac{9.5042z}{z-0.9048}\right]$$

$$= \frac{1}{3}\left[\frac{29.9816z}{z-1} - \frac{39.4858z}{z-0.9753} + \frac{9.5042z}{z-0.9048}\right].$$

Using the table of transforms we get

$$y[i] = \tfrac{1}{3}[29.9816H[i] - 39.4858 \times 0.9753^i + 9.5042 \times 0.9048^i$$

$$= 9.9939H[i] - 13.1619 \times 0.9753^i + 3.1681 \times 0.9048^i.$$

(c) Discrete convolution gives

$$[y] = [G][u].$$

To proceed we need to form the matrix $[G]$ and the vector $[u]$. The latter is easy in this example as all the elements are 0.1 but the former requires the evaluation of a set of values for g_i, namely $g_0, g_1, g_2, \ldots$.

We have found that

$$g_i = \tfrac{10}{3}(0.9753^i - 0.9048^i)$$

thus

$$g_0 = \tfrac{10}{3}(0.9753^0 - 0.9048^0) = 0$$
$$g_1 = \tfrac{10}{3}(0.9753^1 - 0.9048^1) = 0.235$$
$$g_2 = 0.4418$$
$$g_3 = 0.6233$$
$$g_4 = 0.7819$$
$$g_5 = 0.9202$$
$$g_6 = 1.0399$$
$$g_7 = 1.1432$$
$$g_8 = 1.2316$$
$$g_9 = 1.3067.$$

Typically y_0, y_1, y_2, y_3, are obtained from

$$\begin{bmatrix} y_0 \\ y_1 \\ y_2 \\ y_3 \end{bmatrix} = \begin{bmatrix} g_0 & 0 & 0 & 0 \\ g_1 & g_0 & 0 & 0 \\ g_2 & g_1 & g_0 & 0 \\ g_3 & g_2 & g_1 & g_0 \end{bmatrix} \begin{bmatrix} u_0 \\ u_1 \\ u_2 \\ u_3 \end{bmatrix}$$

$$= \begin{bmatrix} 0 & 0 & 0 & 0 \\ 0.2350 & 0 & 0 & 0 \\ 0.4418 & 0.2350 & 0 & 0 \\ 0.6233 & 0.4418 & 0.2350 & 0 \end{bmatrix} \begin{bmatrix} 0.1 \\ 0.1 \\ 0.1 \\ 0.1 \end{bmatrix}.$$

Giving, for example,

$$y_3 = (0.6233 \times 0.1) + (0.4418 \times 0.1) + (0.2350 \times 0.1) + (0 \times 0.1)$$

$$= 0.1300.$$

(d) Finally, we will solve the difference equation using ZT. The relevant equation is in section 5.4

$$4y_{i+2} - 7.5y_{i+1} + 3.51y_i = 0.01 \times 10u_i.$$

Transforming we get, with zero initial conditions

$$4z^2 Y(z) - 7.5z Y(z) + 3.51 Y(z) = 0.1 U(z)$$

from which

$$Y(z) = \frac{0.1 U(z)}{4z^2 - 7.5z + 3.51}.$$

In this case (noting absence of Δ) $U(z) = \dfrac{z}{z-1}$ yielding

$$Y(z) = \frac{0.1z}{(z-1)(4z^2 - 7.5z + 3.51)}.$$

To proceed requires the factorisation of the quadratic, and, using the quadratic formula, gives

$$Y(z) = \frac{2.5 \times 10^{-2} z}{(z-1)(z-0.975)(z-0.9)}.$$

Using partial fractions we get

$$Y(z) = \frac{10z}{z-1} - \frac{13.33z}{z-0.975} + \frac{3.33z}{z-0.9}$$

and then using the table of transforms we get

$$y[i] = [10H[i] - 13.33 \times 0.975^i + 3.33 \times 0.9^i]$$

which we can compare with the result on p. 128. Let us also compare the $Y(z)$ as obtained above with that on p. 127. There we had

$$Y(z) = \frac{1}{3} \left[\frac{z^2}{(z-1)(z-0.9753)} - \frac{z^2}{(z-1)(z-0.9048)} \right]$$

which we can collect up as

$$Y(z) = \frac{2.35 \times 10^{-2} z^2}{(z-1)(z-0.9753)(z-0.9048)}.$$

Once again we notice that the result from the PTF method has z^2 in the numerator and that from the difference equation method has only z. We have met this before.

Let us recapitulate; we have worked the second order problem by a number of different methods which are listed in Table 5.3 together with their calculated values. For the discrete methods the sampling interval is $\Delta = 0.1$ s which is also the interval used in calculating the values from the continuous solution.

Table 5.3 highlights the distinction between results obtained using difference equations formulation and those from the PTF $\leftrightharpoons$ convolution concept which are all different from the

Table 5.3

Continuous Equation	Difference Equation Iteration	Difference Equation ZT	Pulse Transfer Function	Discrete Convolution
0.000 000	0.000 000	0.000 000	0.000 100	0.000 000
0.011 993	0.000 000	0.000 000	0.023 596	0.023 500
0.046 043	0.025 000	0.025 000	0.067 775	0.067 682
0.099 481	0.071 875	0.071 875	0.130 100	0.130 012
0.169 901	0.137 828	0.137 828	0.208 294	0.208 209
0.255 143	0.220 357	0.220 357	0.300 307	0.308 223
0.353 267	0.317 226	0.317 226	0.404 297	0.404 218
0.462 524	0.426 435	0.426 435	0.518 613	0.518 537
0.513 540	0.546 200	0.546 200	0.641 772	0.641 698
0.708 349	0.674 928	0.674 928	0.772 445	0.772 373
0.842 254	0.811 200	0.811 200	0.909 441	0.909 371
0.981 942	0.953 750	0.953 750	1.051 695	1.051 627
1.126 040	1.101 453	1.101 453	1.198 255	1.198 188
1.274 742	1.253 310	1.253 310	1.348 267	1.348 203
1.426 149	1.408 431	1.408 431	1.500 977	1.500 915
1.577 910	1.566 028	1.566 028	1.655 707	1.655 645
1.735 388	1.725 404	1.725 404	1.811 855	1.811 795
1.892 014	1.885 944	1.885 944	1.968 887	1.968 829
2.049 288	2.047 102	2.047 102	2.126 331	2.126 274
2.206 762	2.208 401	2.208 401	2.283 767	2.283 712
2.364 043	2.369 419	2.369 419	2.440 828	

continuous formulation. To appreciate this distinction we must keep in mind that the difference equation formulation is a rather crude approximation to the continuous system whilst the PTF concept is not really attempting to approximate a continuous system but is a concept in its own right for system's with a pulse input*.

5.10   The Zero-Order Hold

Previous sections have identified the mutual consistency of the discrete convolution method and the PTF method. We also noticed that the solutions obtained are the actual values, at the sampling instants, of the output of a continuous system excited by weighted impulses at its input. Fig. 5.13 (p. 116), illustrates this point. To emphasise the point that the two methods calculate y at the sampling instants, we can introduce another sampler at the output, as shown in Fig. 5.14 which is synchronous with that at the input.

Clearly, the box delineated by the broken line encloses a system which can be regarded as discrete with a PTF of $G(z)$. It is excited by pulses and gives a response which is a train of pulses. Obviously we can cascade such elements and form feedback loops just as with continuous systems. The study of such interconnected systems falls into the field of sampled data systems and the reader is referred to texts on control theory which include sampled data studies[2].

Pulses need to be of high amplitude (because of their short duration) to produce enough energy (strength) to yield a noticeable response as excitation in real continuous systems. The result of high amplitude could cause saturation somewhere in the system or, worse, damage.

---

* If backward differences are used different values will be obtained in columns two and three.

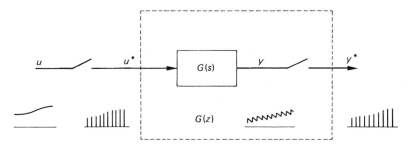

Fig. 5.14

Since many systems, especially control systems, now use computers as controllers their natural output control signals are discrete. This is not compatible with the systems being controlled. To 'interface' a digital controller with a continuous system we can use a 'zero order hold' (ZOH) element. This element is excited by the pulses from the controller and produces a steady output whose magnitude is proportional to the strength of the input pulses and whose duration is that of the sampling interval. Fig. 5.15 shows a ZOH element excited by (a) a single pulse and by (b) a train of pulses the output of which more nearly represents a continuous signal. Note that this type of hold element is known as a zero order hold because it makes use of pulse strength only. There are higher order holds but these will not concern us here.

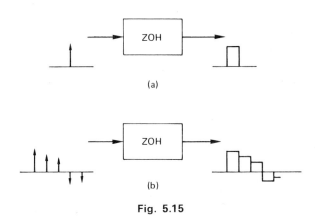

Fig. 5.15

An input impulse $u(t_0).\Delta.\delta(t)$ will produce an output pulse height $u(t_0)$ and duration Δ which can be represented by a step of height $u(t_0)$ at $t=0$ followed by a negative step of the same height at $t=\Delta$, i.e. $(H(t)-H(t-\Delta))u(t_0)$. Thus, the impulse response of the ZOH is $\frac{1}{\Delta}[H(t)-H(t-\Delta)]$ and its TF is

$$\frac{1}{\Delta}\left(\frac{1}{s}-\frac{1}{s}e^{-s\Delta}\right)=\frac{1-e^{-s\Delta}}{\Delta s}.$$

We will illustrate the effect of a ZOH by preceding a first order system with one as shown in Fig. 5.16.

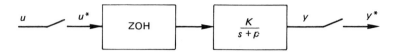

Fig. 5.16

The overall TF is

$$G(s) = \left(\frac{1-e^{-s\Delta}}{\Delta s}\right)\left(\frac{K}{s+p}\right)$$

$$= \frac{K}{\Delta s(s+p)} - \frac{Ke^{-s\Delta}}{\Delta s(s+p)}.$$

We can find the ZT of the first term on the right quite easily using a table of transforms.

$$\frac{K}{\Delta s(s+p)} \rightarrow \frac{K(1-e^{-p\Delta})z}{\Delta p(z-1)(z-e^{-p\Delta})}$$

The second term is the same except that it is delayed by one interval, thus the above ZT only needs to be multiplied by z^{-1}, giving

$$\frac{Ke^{-s\Delta}}{\Delta s(s+p)} \rightarrow \frac{K(1-e^{-p\Delta})}{\Delta p(z-1)(z-e^{-p\Delta})}.$$

On combining the two components of $G(z)$ we get

$$G(z) = \frac{K(1-e^{-p\Delta})}{\Delta p(z-e^{-p\Delta})}.$$

Let u be a unit step so that $U(z) = \dfrac{z}{z-1}\Delta$ and then

$$Y(z) = \frac{K(1-e^{-p\Delta})z}{p(z-1)(z-e^{-p\Delta})}.$$

Using the table again we find the time solution to be

$$y[i] = \frac{K}{p}(1-e^{-pi\Delta}), \qquad i \geqslant 0.$$

Remembering that $i\Delta = t$ we notice that this is the exact time response for a continuous system, but a moment's thought tells us that the effect of the ZOH in this case, which has a 'pulse-step' input, is to convert the train of pulses back into a 'continuous-step' as shown in Fig. 5.17. For comparison let us re-work this without the ZOH. Then

$$G(z) = \frac{Kz}{z-e^{-p\Delta}}$$

and

$$Y(z) = \frac{Kz^2\Delta}{(z-1)(z-e^{-p\Delta})}$$

or

$$Y(z) = \frac{Kz\Delta}{(z-1)(z-e^{-p\Delta})} \cdot z.$$

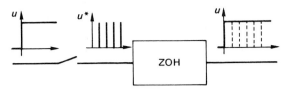

Fig. 5.17

From this we recognise this as being the output from a system $\dfrac{Kz\Delta}{(z-1)(z-e^{-p\Delta})}$ advanced by one interval

$$\frac{Kz\Delta}{(z-1)(z-e^{-p\Delta})} \rightarrow \frac{K\Delta}{(1-e^{-p\Delta})}(1-e^{-pi\Delta})$$

$$\therefore y[i] = \frac{K\Delta}{(1-e^{-p\Delta})}(1-e^{-(i+1)\Delta}) \qquad i \geqslant 0.$$

If we let $K=2$, $p=1$, $u=10$ and $\Delta=0.1$ we can get numerical results to compare with our previous first order example, which are shown in Table 5.4.

Table 5.4

With ZOH	Without ZOH
0.000	2.00
1.903	3.81
3.625	5.40
5.184	6.93
6.594	8.27

A physical insight into the effect of a ZOH can be obtained by looking at the impulse response of the combination of ZOH and a first order lag, (Fig. 5.18). The ZOH converts the unit impulse input into a pulse of height 1 and duration Δ. This is then the input to the lag which sees the leading edge of the pulse as a unit step and responds with an exponential rise. The trailing edge of the pulse is seen as a negative step and the lag now gives a decaying exponential output from the level reached at time Δ. It is easy to compute the overall impulse response of the ZOH and first order lag and we get,

$$g_0 = 0$$
$$g_1 = 1.9033$$
$$g_2 = 1.7220$$
$$g_3 = 1.5580$$
$$g_4 = 1.4100$$
$$g_5 = 1.2760$$
$$g_6 = 1.1540$$
$$g_7 = 1.0440$$
$$g_8 = 0.9450$$
$$g_9 = 0.8554$$
$$g_{10} = 0.7740.$$

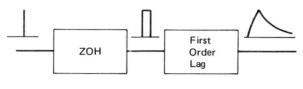

Fig. 5.18

Using these in convolution with a step input of 10 units gives

$$y_0 = 0$$
$$y_1 = 1.903$$
$$y_2 = 3.625$$
$$y_3 = 5.183$$

which are identical with the 'continuous' results and it follows that a better discrete approximation to a continuous system is obtained when ZOH is included.

5.11 Interpretation of Transfer Functions

In section 5.1 we remarked that an important aspect of an engineer's work is the prediction of the performance of some system or other. If a complete quantitative prediction is required (rather rarely in routine engineering practice but often in design and in the examination room!) then it is necessary to invoke one of the computational methods outlined above and, of course, its value and accuracy will depend on the accuracy by which the system parameters, such as gains and time constants, are known. Indeed it may be a waste of time and effort persuing complicated computations using unreliable parameters, and further, it may well be dangerous to flaunt the results of such efforts because they may unreliably be used for decision making.

However, it is often only necessary to assess or estimate a prediction and this can be done very easily by inspecting the system transfer function, $G(s)$ or $G(z)$ as appropriate. Often modelling operations naturally throw up a transfer function in a standard form. (When feedback exists some manipulations may be necessary to get a standard form but these are a subject in control theory.[2,14])

Suppose we have arrived at a TF in the form

$$G(s) = \frac{K(s+2)}{(s+1)(s+3)}.$$

We know that this can be written as

$$G(s) = \frac{A}{s+1} + \frac{B}{s+3}$$

from which

$$g(t) = Ae^{-t} + Be^{-3t}$$

and that the characteristic response of the system contains two exponential decays with time constants 1 s and 1/3 s. If we had

$$G(s) = \frac{K(s+2)}{(s-1)((s+3)^2 + 25)}$$

we would know that one term of the characteristic response was an exponential increase of time constant 1s and the other term was a damped sine wave with damping time constant of 1/3 sec. and a frequency of $\sqrt{25}=5$ rad s^{-1}. Incidentally, this system would be unstable because of the exponential increase.

It follows that the characteristic response of a system can be estimated (in form) by the nature of the a's and b's etc. in the TF. Table 5.5 classifies various possibilities.

Obviously similar conclusions can be drawn from the PTF and Table 5.6 also classify these.

The nature of the a's and b's can be represented graphically and these ideas are developed now. Remember that both s and z are complex numbers making $G(s)$ and $G(z)$ functions of a complex variable which are in themselves complex numbers. Bearing this in mind we see that if we choose a value of s equal to the negative of say a, i.e. $s=-a$ then the function $G(s)$ would take on an infinite value. This would also be the case for the other value of b, whether they were real or complex. For this reason we call $-a$, $-b$, etc. the poles of the TF. Using the same idea we call -2 in the above example the zero of the TF. In a similar way we can identify the poles and zeros of $G(z)$; a PTF.

We can plot the position of the poles and zeros on an Argand diagram, that for $G(s)$ is known as the s-plane and that for $G(z)$ is the z-plane. See Fig. 5.19. Pole-zero maps can be used as a means of graphically calculating the value of a transfer function for a given value of s (or z as relevant) as follows. Since a TF is a function of the complex variable s, it is, itself, a complex number having a modulus $|G(s)|$ and argument $\underline{/G(s)}$. Both these quantities can be determined from the pole zero map.

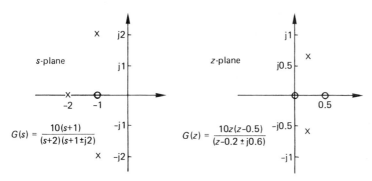

Fig. 5.19

Consider the simplest form of TF, namely

$$G(s)=\frac{K}{s+p}.$$

Then
$$|G(s)|=\frac{K}{|s+p|}\qquad\text{and}\qquad \underline{/G(s)}=-\underline{/s+p}.$$

Inspecting its pole-zero diagram (Fig. 5.20) we see that for some arbitrary value of s, $\overrightarrow{(OQ)}$, the phasor $\overrightarrow{PQ}$ is $s+p$, thus $PQ=|s+p|$ and $\underline{/QPO}=\underline{/s+p}$ from which

$$|G(s)|=\frac{K}{(PQ)}\qquad\text{and}\qquad \underline{/G(s)}=-\underline{/QPO}\ .$$

Table 5.5

$G(s)$	s-plane	$g(t)$	Impulse Response
$\dfrac{K}{s+a}$	(pole at $-a$)	Ke^{-at}	
$\dfrac{K}{s-a}$	(pole at $+a$)	Ke^{at}	
$\dfrac{K}{(s+a)(s+b)} \equiv \dfrac{K}{s^2 + 2\zeta\omega_k s + \omega_n^2}$ $\zeta > 1$	(poles at $-b$, $-a$)	$\dfrac{K}{b-a}(e^{-at} - e^{-bt})$	

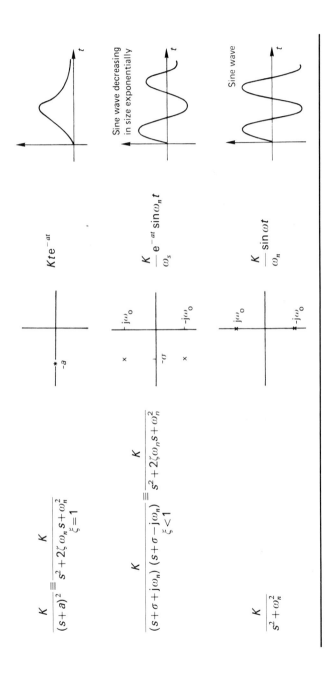

$$\frac{K}{(s+a)^2} \equiv \frac{K}{s^2 + 2\zeta\omega_n s + \omega_n^2}\Big|_{\zeta=1}$$

$$Kte^{-at}$$

$$\frac{K}{(s+\sigma+j\omega_n)(s+\sigma-j\omega_n)} \equiv \frac{K}{s^2 + 2\zeta\omega_n s + \omega_n^2}\Big|_{\zeta<1}$$

$$\frac{K}{\omega_s}e^{-at}\sin\omega_n t$$

Sine wave decreasing in size exponentially

$$\frac{K}{s^2 + \omega_n^2}$$

$$\frac{K}{\omega_n}\sin\omega t$$

Sine wave

Table 5.6

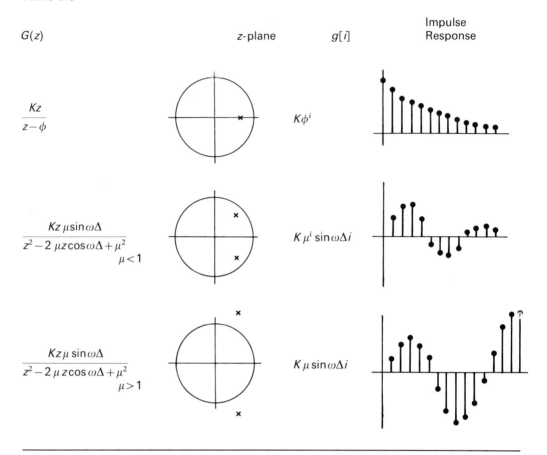

$G(z)$	z-plane	$g[i]$	Impulse Response
$\dfrac{Kz}{z-\phi}$		$K\phi^i$	
$\dfrac{Kz\,\mu\sin\omega\Delta}{z^2-2\,\mu z\cos\omega\Delta+\mu^2}$ $\mu<1$		$K\mu^i\sin\omega\Delta i$	
$\dfrac{Kz\mu\sin\omega\Delta}{z^2-2\,\mu z\cos\omega\Delta+\mu^2}$ $\mu>1$		$K\mu\sin\omega\Delta i$	

Clearly this evaluation could be repeated for any value of s at any point on the plane. We shall see later that a certain range for s is of special interest. For more complicated TFs the idea is extended as shown in Fig. 5.21 for the TF

$$G(s)=\frac{K(s+z)}{(s+p_1)(s+p_2)}$$

in which p_1 and p_2 are a complex conjugate pair.

$$|G(s)|=\frac{K(ZQ)}{(P_1Q)(P_2Q)}$$

$$\underline{/G(s)}= \underline{/QZO} - \underline{/QP_1O'} - \underline{/QP_2O''}.$$

Obviously we can use these ideas on the z-plane for PTFs.

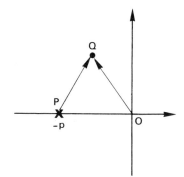

Fig. 5.20

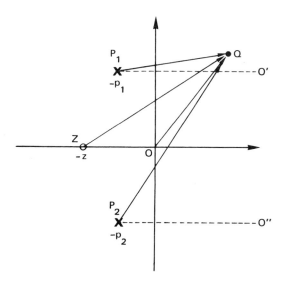

Fig. 5.21

5.12 Stability

Stability was mentioned in passing in the last section but now we will look a little more closely at the subject in relation to signal processing. It is a vast subject and reference to texts on control theory[2,14] will provide considerable detail, however, this extensive coverage is not within our scope although signal processing methods are relevant to it.

There are many definitions of stability but for our purposes we will define it as follows. A system is stable if for a bounded input there is a bounded output (a BIBO system). This means that if the input signal to a system has bounds on its amplitude, that is, the signal does not extend to infinite amplitude, then it is stable if the output remains bounded.

From Table 5.5 we deduce that if a system has a TF, $G(s)$ with a p which is real and negative or is one of a complex pair with negative real parts, then a bounded input will produce an output with a component which increases indefinitely meaning that the system is unstable. Such a system would have poles on the r.h.s. of the s-plane.

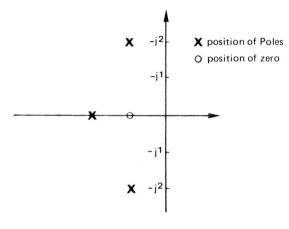

Fig. 5.22

The third order system, Fig. 5.19 used as an example in the previous section produces a pole-zero map in the s-plane as shown in Fig. 5.22 and since it is stable the poles are on the l.h.s. of the plane. Clearly, if the form of the TF has a factored denominator stability assessment is straightforward but, as we shall see, this is not always the case.

Similarly, with a PTF we examine the nature of the poles to draw a conclusion on stability, Table 5.2. A simple PTF component with a single real pole produces a ZT pair

$$\phi^i \rightleftharpoons \frac{z}{z-\phi}.$$

If $|\phi| < 1$ the term ϕ^i decreases as i increases and if $|\phi| > 1$ the reverse occurs. If $|\phi| = 1$ $\phi^i = 1$.

The behaviour resulting from a complex conjugate pair of poles can be deduced by considering the following. If we take the ZT pair

$$e^{-ip\Delta} \sin \omega i\Delta \rightleftharpoons \frac{ze^{-p\Delta}\sqrt{1-\cos^2 \omega\Delta}}{z^2 - 2ze^{-p\Delta}\cos \omega\Delta + e^{-2p\Delta}}$$

and let $e^{-p\Delta} = \mu$ and $\cos \omega\Delta = \gamma$ we can put

$$\mu^i \sin \omega i\Delta \rightleftharpoons \frac{z\mu\sqrt{1-\gamma^2}}{z^2 - 2\mu\gamma z + \mu^2} \qquad \text{for } \gamma < 1.$$

Now if $\mu < 1$, $\mu^i \sin \omega i\Delta$ decreases in amplitude as i increases and vice versa if $\mu > 1$, we notice that since the two poles are

$$\mu\gamma \pm j\mu\sqrt{1-\gamma^2} = \mu(\gamma \pm j\sqrt{1-\gamma^2})$$
$$= \mu(\cos \omega\Delta \pm j\sin \omega\Delta)$$

and since $\qquad |\cos \omega\Delta \pm j\sin \omega\Delta| = 1$

then $\qquad\qquad\qquad \mu = |\mu\gamma \pm j\mu\sqrt{1-\gamma^2}|.$

All this tells us that if a system has a PTF, $G(z)$, with a ϕ which is real and for which $|\phi|$ is greater than unity or is one of a complex pair for which $|\phi|$ is greater than unity then the

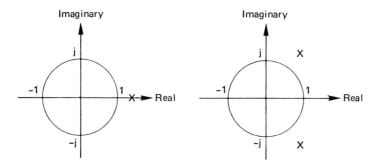

Fig. 5.23

system is unstable. Such a system would have poles outside a unit circle on the *z*-plane, see Fig. 5.23.

All this seems straightforward but we must bear in mind a difficulty. The ideas above all assume that we know the denominator of the TF or the PTF in factored form but this is often not the case, especially with closed loop systems. We illustrate the difficulty with an example.

For Fig. 5.24 the closed loop TF is

$$\frac{10(s+1)}{(s+2)(s^2+2s+5)+10(s+1)} = \frac{10(s+1)}{s^3+4s^2+19s+20}.$$

The same problem arises in closed loop PTFs.

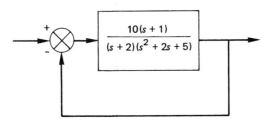

Fig. 5.24

Now, before we can assess stability we need to factorise the third order denominator but since there is no straightforward method we have to resort to special techniques. Elegant though these methods are, with names such as root locus, Nyquist plots, Bode plots, we must refrain from following them as they are in the realm of control theory which is not within the scope of this book. The reader is recommended to the many excellent texts on the subject.[2]

However, to highlight two problems regarding the stability of feedback systems we will look at a simple example. We will investigate the system illustrated in Fig. 5.25. *K* is a gain parameter and we will investigate the system behaviour for different values. Using the methods already developed we can find $G(z)$ to be

$$G(z) = \frac{K(1 - e^{-\Delta})}{(z - e^{-\Delta})}.$$

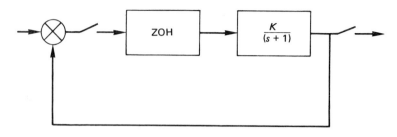

Fig. 5.25

Then the closed loop TF becomes

$$\frac{K(1-e^{-\Delta})}{\{z-[(K+1)e^{-\Delta}-K]\}}.$$

If we take $\Delta = 0.1$ as before this becomes

$$\frac{0.0952\,K}{\{z-[(K+1)0.9048-K]\}}.$$

Taking $\phi = (K+1)0.9048 - K$ and using a range of values of K we obtain the figures in Table 5.7. We notice that as K increases, ϕ changes from positive values through zero to negative values and eventually to values giving $|\phi| > 1$. Thus as K increases the system will become unstable and the value of K which just produces instability is

when
$$|(K+1)0.9048 - K| = 1$$

i.e.
$$K = 20.0084.$$

This effect (increasing K causing instability) is also relevant to continuous systems but is only true in that case for third order systems and higher order systems. A prototype continuous second order system in a feedback loop cannot become unstable unless the damping factor is negative; this is practically an artificial condition.

The second problem involves the choice of Δ. Let us take $K = 10$ and consider different values of Δ, then $\phi = 11e^{-\Delta} - 10$. Values of ϕ are tabulated for a range of values of Δ in

Table 5.7

K	ϕ
0	0.9048
1	0.8097
2	0.7145
5	0.4290
7	0.2387
10	−0.0468
15	−0.5226
20	−0.9984
25	−1.474
30	−1.95

Table 5.8

Δ	φ
0.05	0.4635
0.1	−0.0468
0.15	−0.5322
0.2	−0.9940
0.25	−1.4332
0.3	−1.8510

Table 5.8. Again an increase in Δ causes instability and the system is just unstable when

$$|11e^{-\Delta} - 10| = 1$$

giving $$\Delta = 0.200\,67.$$

Of course, this effect is not relevant to continuous systems.

Clearly we must take care when designing feedback systems because K (and Δ for discrete systems) are important design parameters as far as stability is concerned.

5.13 Random Signals in Linear Systems: Cross-correlation

So far we have considered input–output signal processing methods in which we have implicitly assumed the input signal to be deterministic or in the form of data made up of signal samples. Continuous convolution requires a deterministic signal and discrete convolution requires discrete data. In the case of a random signal we certainly do not have information in the form required for continuous processing. We could, however, take samples of a random signal to effect discrete processing but the results would not tell us much. Thus we approach the problem of random signals differently, as we did in signal characterisation.

Recall that the cross-correlation function relating two signals is a characterisation function. Here we are going to deduce the cross-correlation function relating the input and output signals of a linear system. Clearly the relationship will depend, not only on the input signal, but also on the system itself. Indeed we shall find that the cross-correlation function characterises the system itself under certain conditions.

We have

$$R_{uy}(\tau) = \underset{T \to \infty}{\text{Limit}} \frac{1}{2T} \int_{-T}^{T} u(t) . y(t + \tau) \, dt \qquad (5.19)$$

and

$$y(t) = \int_{0}^{\infty} g(\eta) \, u(t - \eta) \, d\eta \qquad (5.20)$$

and

$$y(t + \tau) = \int_{0}^{\infty} g(\eta) \, u(t + \tau - \eta) \, d\eta. \qquad (5.21)$$

Remember that our studies of random signals are only relevant for steady-state conditions. Thus we need the appropriate form of the convolution integral (5.20).

Now substituting equation (5.21) into (5.19) we get

$$R_{uy}(\tau) = \underset{T \to \infty}{\text{Limit}} \frac{1}{2T} \int_{-T}^{T} u(t) \int_{0}^{\infty} g(\eta) \cdot u(t + \tau - \eta) \, d\eta \, dt.$$

This is a double integral and we can reduce it to a single integral by changing the order of integration

$$R_{uy}(\tau) = \int_{0}^{\infty} g(\eta) \left[\underset{T \to \infty}{\text{Limit}} \frac{1}{2T} \int_{-T}^{T} u(t) \cdot u(t + \tau - \eta) \, dt \right] d\eta.$$

We can now recognise the term within brackets above as the defining integral for $R_{uu}(\tau - \eta)$, thus

$$R_{uy}(\tau) = \int_{0}^{\infty} g(\eta) R_{uu}(\tau - \eta) \, d\eta. \tag{5.22}$$

This is a formula which enables us to find the input–output cross-correlation when we know the system impulse response and the auto-correlation of the input signal. Note that it is a convolution integral.

We will illustrate its use by means of the example shown in Fig. 5.26. Substituting into equation (5.22) we get

$$R_{uy}(\tau) = \int_{0}^{\infty} 2e^{-2\eta} \cdot 4 \cdot e^{-3|\tau - \eta|} d\eta$$

$$= 8 \int_{0}^{\infty} e^{-2\eta} e^{-3|\tau - \eta|} d\eta.$$

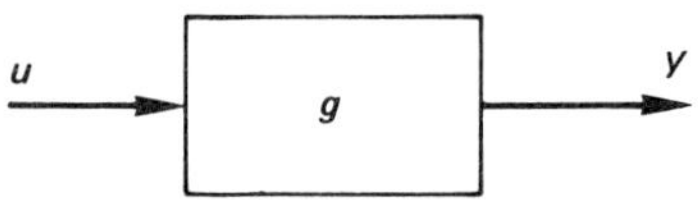

$R_{uu}(\lambda) = 4e^{-3|\lambda|}, \quad g(\eta) = 2e^{-2\eta}$

Find $R_{uy}(\tau)$

Fig. 5.26

Since $R_{uu}(\ )$ is an even function made up of a decaying exponential and its mirror image, we need to look carefully at its form and the ranges of integration. Fig. 5.27 illustrates this with η as abscissa. Because of the discontinuity at $\eta = \tau$ the range of integration 0 to ∞ must be divided into two, namely 0 to τ and τ to ∞, thus

$$R_{uy}(\tau) = 8 \left[\int_{0}^{\tau} e^{-2\eta} e^{-3(\tau - \eta)} \, d\eta + \int_{\tau}^{\infty} e^{-2\eta} e^{3(\tau - \eta)} \, d\eta \right]$$

$$= 8 \left[e^{-3\tau} \int_{0}^{\tau} e^{\eta} \, d\eta + e^{3\tau} \int_{\tau}^{\infty} e^{-5\eta} \, d\eta \right]$$

$$= 8 \left[e^{-3\tau} (e^{\eta})_{0}^{\tau} + e^{3\tau} \left(\frac{e^{-5\eta}}{-5} \right)_{\tau}^{\infty} \right]$$

$$= 8 \left[e^{-3\tau} (e^{\tau} - 1) + \frac{e^{3\tau}}{5} (0 + e^{-5\tau}) \right]$$

$$= 9\tfrac{3}{5} e^{-2\tau} - 8 e^{-3\tau}. \tag{5.23}$$

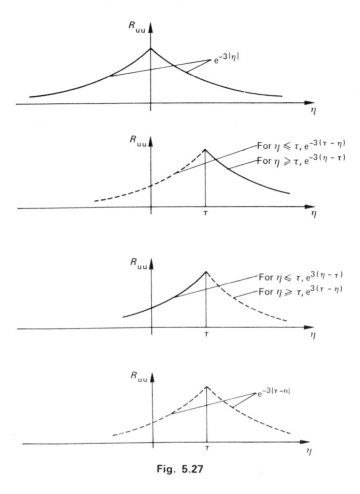

Fig. 5.27

This result is valid for $\tau \geqslant 0$ (see Fig. 5.27), however τ can be less than zero in which case $R_{uu}(\ )$ appears as in Fig. 5.28. Now only $e^{3(\tau - \eta)}$ is involved in the integration and the range is continuous from 0 to ∞, thus

$$R_{uy}(\tau) = 8 \int_0^\infty e^{-2\eta} e^{3(\tau - \eta)} d\eta$$

$$= 8e^{3\tau} \int_0^\infty e^{-5\eta} d\eta$$

$$= \tfrac{8}{5} e^{3\tau} \qquad \text{for } \tau \leqslant 0. \qquad (5.24)$$

Observe that when τ is zero in both equations (5.23) and (5.24) we get

$$R_{uy}(0) = \tfrac{8}{5}$$

$R_{uy}(\tau)$ is illustrated in Fig. 5.29.

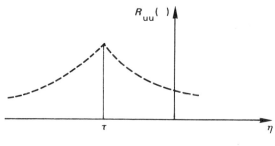

Fig. 5.28

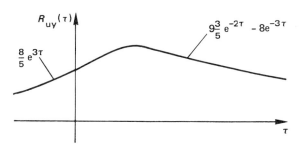

Fig. 5.29

An important and very interesting result arises when the input u is white noise (see p. 202) for which $R_{uu}(\tau)$ is an impulse function $R_{uu}(\tau) = N\delta(\tau)$ where N is a weighting constant. Then

$$R_{uy}(\tau) = \int_0^\infty g(\eta) . N . \delta(\tau - \eta) \mathrm{d}\eta. \tag{5.25}$$

Recall that $\delta(\tau - \eta)$ is zero except when $\tau - \eta = 0$, that is, when $\eta = \tau$, thus $g(\eta)$ can only contribute to the integrand when $\eta = \tau$. The only relevant value for $g(\eta)$ is $g(\tau)$ which is now constant as far as the integration is concerned, then

$$R_{uy}(\tau) = Ng(\tau) \int_0^\infty \delta(\tau - \eta) \mathrm{d}\eta$$

$$= Ng(\tau) \tag{5.26}$$

since $\int_0^\infty \delta(\tau - \eta) \mathrm{d}\eta = 1$ from the definition of the impulse function.

The significance of equation (5.26) is that if a system is excited by white noise the input–output cross-correlation function is proportional to the inpulse response. This important result will be discussed further in the chapter on case studies.

A word or two here is necessary to discuss the constant N. Clearly $R_{uu}(\ )$ is dimensionally (signal × quantity)^2 and since $\delta(\ )$ is dimensionally (Time)$^{-1}$ we deduce that N is dimensionally (signal × quantity)^2 × time and represents the strength (area) of the impulse $R_{uu}(\tau) = N\delta(\tau)$. We shall meet N again later.

Since equation (5.22) is a convolution integral we could, in principle, invoke discrete convolution or its matrix equivalent. However, the upper limit is infinite and thus necessitates an infinite number of terms in the convolution sum or, which amounts to the same thing, requires matrices of infinite order. Clearly the practical implementation is unwieldly.

5.14   Random Signals in Linear Systems: Auto-correlation

Recall from Chapter 3 that auto-correlation is a characterisation function of a signal and it is reasonable to want to evaluate it for the output of a system when we know the characterisation of the input and of the system. We look for a relationship between input auto-correlation, system impulse response and output auto-correlation.
We have

$$R_{yy}(\tau) = \underset{T \to \infty}{\text{Limit}} \frac{1}{2T} \int_{-T}^{T} y(t) \cdot y(t + \tau) dt \qquad (5.27)$$

but

$$y(t) = \int_{0}^{\infty} g(\eta) u(t - \eta) d\eta \qquad (5.28)$$

and

$$y(t + \tau) = \int_{0}^{\infty} g(\lambda) \cdot u(t + \tau - \lambda) d\lambda. \qquad (5.29)$$

We need the two dummy variables η and λ in order to keep tabs on the separate integrations and, as before, we need the steady state form of the convolution integrals.
   Substituting equations (5.28) and (5.29) into (5.27) we get

$$R_{yy}(\tau) = \underset{T \to \infty}{\text{Limit}} \frac{1}{2T} \int_{-T}^{T} \left[\int_{0}^{\infty} g(\eta) u(t - \eta) d\eta \int_{0}^{\infty} g(\lambda) u(t + \tau - \lambda) d\lambda \right] dt.$$

Using the same device as in the previous section we can reduce this triple integral into a double integral, that is, change the order of integration and recognise the defining expression for input auto-correlation.

$$R_{yy}(\tau) = \int_{0}^{\infty} \int_{0}^{\infty} g(\eta) \cdot g(\lambda) \left[\underset{T \to \infty}{\text{Limit}} \frac{1}{2T} \int_{-T}^{T} u(t - \eta) u(t + \tau - \lambda) dt \right] d\lambda d\eta$$

$$= \int_{0}^{\infty} \int_{0}^{\infty} g(\eta) g(\lambda) R_{uu}(\tau - \lambda + \eta) d\lambda d\eta. \qquad (5.30)$$

This is the formula we require and it allows us to calculate $R_{yy}(\ )$ when we know $g(\ )$ and $R_{uu}(\ )$.
   To illustrate we will use the same system as the previous example (Fig. 5.26) but now we want to find $R_{yy}(\tau)$. Inspecting equation (5.30) we see that we can integrate with respect to λ firstly, that is we need to integrate

$$\int_{0}^{\infty} g(\lambda) R_{uu}(\tau - \lambda + \eta) d\lambda.$$

If we put $\tau + \eta = k$ we get

$$\int_{0}^{\infty} g(\lambda) R_{uu}(k - \lambda) d\lambda$$

which is the same form as equation (5.22). On substituting the relevant terms, that is,

$g(\lambda)=2e^{-2\lambda}$ and $R_{uu}(k-\lambda)=4e^{-3|k-\lambda|}$ we get

$$8\int_0^\infty e^{-2\lambda}e^{-3|k-\lambda|}d\lambda=9\tfrac{3}{5}e^{-2k}-8e^{-3k} \qquad \text{for } k\geqslant0$$

and

$$8\int_0^\infty e^{-2\lambda}e^{3(k-\lambda)}d\lambda=\tfrac{8}{5}e^{3k} \qquad \text{for } k\leqslant0.$$

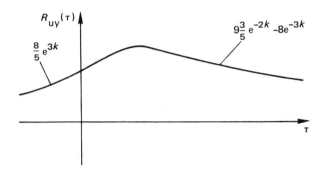

Fig. 5.30

This function is illustrated in Fig. 5.29 but is reproduced here as Fig. 5.30. To complete the integration process we now need to evaluate

$$R_{yy}(\tau)=\int_0^\infty 2e^{-2\eta}R(k)d\eta$$

in which
$$R(k)=9\tfrac{3}{5}e^{-2k}-8e^{-3k} \qquad \text{for } k\geqslant0$$
and
$$=\tfrac{8}{5}e^{3k} \qquad \text{for } k\leqslant0.$$

This becomes
$$R(\tau+\eta)=9\tfrac{3}{5}e^{-2(\tau+\eta)}-8e^{-3(\tau+\eta)} \qquad \text{for } \tau+\eta\geqslant0$$
$$=\tfrac{8}{5}e^{3(\tau+\eta)} \qquad \text{for } \tau+\eta\leqslant0.$$

The range of integration on η is 0 to ∞, that is, the range is such that η is only positive.

also $\quad \tau+\eta\geqslant0 \quad$ implies $\quad \eta\geqslant-\tau$
and $\quad \tau+\eta\leqslant0 \quad$ implies $\quad \eta\leqslant-\tau$

all of which will be needed in sorting out $R_{yy}(\tau)$ in this example. When τ is positive $R(\tau+\eta)$, on η as abscissa will be as shown in Fig. 5.31.

Then $\qquad R_{yy}(\tau)=\int_0^\infty 2e^{-2\eta}[9\tfrac{3}{5}e^{-2(\tau+\eta)}-8e^{-3(\tau+\eta)}]d\eta$

$$=2\left[9\tfrac{3}{5}e^{-2\tau}\int_0^\infty e^{-4\eta}d\eta-8e^{-3\tau}\int_0^\infty e^{-5\eta}d\eta\right]$$

$$=\tfrac{24}{5}e^{-2\tau}-\tfrac{16}{5}e^{-3\tau} \qquad \text{for } \tau\geqslant0$$

when τ is negative $R(\tau+\eta)$, on η as abscissa, will be as shown in Fig. 5.32.

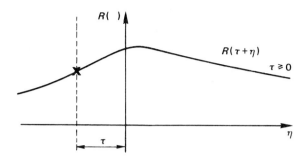

Fig. 5.31

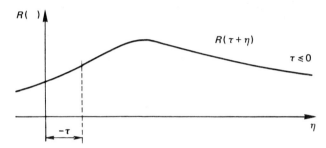

Fig. 5.32

Then $\displaystyle R_{yy}(\tau) = \int_0^{-\tau} 2e^{-2\eta} \cdot \tfrac{8}{5} e^{3(\tau+\eta)} \,d\eta$

$$= + \int_{-\tau}^{\infty} 2e^{-2\eta}(9\tfrac{3}{5}e^{-2(\tau+\eta)} - 8e^{-3(\tau+\eta)})\,d\eta$$

$$= \tfrac{16}{5}e^{3\tau} \int_0^{-\tau} e^{\eta}\,d\eta + \tfrac{96}{5}e^{-2\tau}\int_{-\tau}^{\infty} e^{-4\eta}\,d\eta - 16e^{-3\tau}\int_0^{\infty} e^{-5\eta}\,d\eta$$

$$= \tfrac{16}{5}[e^{2\tau} - e^{3\tau}] + \tfrac{24}{5}e^{2\tau} - \tfrac{16}{5}e^{2\tau}$$

$$= \tfrac{24}{5}e^{2\tau} - \tfrac{16}{5}e^{3\tau} \qquad \text{for } \tau \leqslant 0.$$

As expected the result for $R_{yy}(\tau) = R_{yy}(-\tau)$ is

$$R_{yy}(\tau) = \tfrac{24}{5}e^{-2|\tau|} - \tfrac{16}{5}e^{-3|\tau|}$$

see Fig. 5.33.

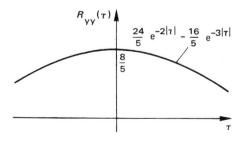

Fig. 5.33

It is obvious from the previous examples that time domain solutions are usually tedious, if not difficult to apply. We shall find that frequency domain methods are far easier but time domain methods do give an insight into the behaviour of systems and characterisations which are not apparent in the other methods. The practical implementation of the discrete method in the time domain also suffers from computational complexity. However, the advent of sophisticated computer algorithms does allow many time domain methods to be used.

Exercises 5

1   A first order system has an impulse response given by

$$g(t) = 4e^{-2t} \qquad \text{for } t \geq 0$$
$$= 0 \qquad\qquad t < 0$$

and is excited by an aperiodic signal of the form

$$u(t) = 2e^{-4t} \qquad \text{for } t \geq 0$$
$$= 0 \qquad\qquad \text{for } t < 0.$$

Using both the Laplace Transform method and the convolution method show that the output is given by

$$g(t) = 4[e^{-2t} - e^{-4t}].$$

Use the final value theorem of the Laplace Transform and the appropriate limit in the convolution integral to find the steady state value of the output.

2   The first order system whose impulse response is

$$g(t) = 4e^{-2t} \qquad \text{for } t \geq 0$$
$$= 0 \qquad\qquad \text{for } t < 0$$

is excited by the cosine signal $u(t) = 5 \cos 2t$ which begins at $t = t_0$. By both the Laplace Transform method and the convolution integral method find the output signal $y(t)$ for the cases when (a) $t_0 = 0$ and (b) $t_0 = -\infty$.
(Hint: Let $5 \cos 2t = \frac{5}{2}(e^{j2t} + e^{-j2t})$ in the convolution integral).

3   A second order system has an impulse response given by

$$g(t) = \frac{1}{2}(e^{-2t} - e^{-4t}) \qquad \text{for } t \geq 0$$
$$= 0 \qquad\qquad\qquad \text{for } t < 0.$$

It is excited by a single rectangular pulse of height 2 units and extending from $t = -\tau$ to $t = +\tau$.
Find the system output $y(t)$.
(Hint: 1. Convolution method. Form the convolution integral for three ranges of t, (a) $t < -\tau$, (b) $-\tau < t < \tau$, (c) $t > \tau$.
Evaluate each one separately, the result of each will be valid for each range. The complete response will be the combination of each piecewise response.

2. Laplace Transform method. Shift the time origin so that the impulse begins at $t=0$ and ends at $t=2\tau$. The pulse can then be regarded as two steps, a positive going one beginning at $t=\tau$ followed by a negative going one at $t=2\tau$. The response is again piecewise.

4 The first order system whose impulse response is

$$g(t)=4e^{-2t} \qquad \text{for } t \geqslant 0$$

$$=0 \qquad \text{for } t<0$$

is excited by a signal given by

$$u(t)=2(1-e^{-2at})$$

show that the response is

$$y(t)=4+\frac{8}{18}e^{-2at}-\frac{80}{18}e^{-2t}.$$

5 Use a sampling interval of $\Delta=0.05$ and form $G(z)$ and $U(z)$ for the system and input signal set out in the previous problem. Then show that the output is given by

$$y_i \simeq 4.19+0.264\,e^{-i}-4.46\,e^{-0.1i}.$$

6 Find the pulse response function corresponding to

$$g(t)=4\,e^{-2t} \qquad \text{for } t \geqslant 0$$

$$=0 \qquad \text{for } t<0 \qquad \text{for } \Delta=0.05$$

and u_i corresponding to $u(t)=2(1-e^{-2at})$. Then use discrete convolution to find the first few values of the output y_i. Compare these with the values obtained from the results in questions 4 and 5.

7 The system and the input of question 4 can be represented by the differential equation

$$\frac{dy}{dx}+2y=8(1-e^{-2at}).$$

Prove this. Show that this can be transformed into the difference equation given by

$$y_{i+1}-0.9y_i=0.4(1-e^{-i}).$$

(a) Find an iterative solution, i.e. use the iterative method to calculate the first four values of y_i.

(b) Using z transforms show that the solution is

$$y_i \simeq 4-(4.758 \times 0.9^i)+(0.752 \times 0.368^i)$$

and calculate the first few values.

8 The first order system in question 4 is preceeded by a zero-order-hold and the input applied to the latter is the same signal as before but sampled every 0.05s. Show that the sampled output of the system is

$$y \simeq 4.005-(4.715 \times 0.9048^i)+(0.710 \times 0.3679^i)$$

calculate the first few values of y_i.

9 Examine the poles of the following transfer functions and make a decision on stability.

(a) $\dfrac{k(s+2)}{s^2-2s-3}$ (b) $\dfrac{k(s+2)}{s^2+2s-3}$

(c) $\dfrac{k(s+2)}{s^2+4s+3}$ (d) $\dfrac{k}{s^2+4^s+13}$

(e) $\dfrac{k}{s^2-45+13}$ (f) $\dfrac{kz}{z^2-1.7z+0.71}$

(g) $\dfrac{kz}{z^2+1.4z-0.72}$ (h) $\dfrac{kz}{z^2-0.5z+0.25}$

(i) $\dfrac{kz}{z^2-1.5z+2.25}$ (j) $\dfrac{kz}{z^3+1.5z^2-0.75z+0.5}$

Hint: In (j) the denominator can be factorised. Choose a factor of the form, $(z\pm a)$ and divide it into the polynomial—trial and error.

10 The system whose impulse response is

$$g(t)=\tfrac{1}{2}(e^{-2t}-e^{-4t}) \qquad \text{for } t\geqslant 0$$
$$=0 \qquad \text{for } t<0$$

has an input signal which is random with an autocorrelation function given by

$$R_{uu}(\tau)=10\delta(\tau).$$

Find the input–output cross-correlation function $R_{uy}(\tau)$ and sketch.

11 The system described in question 10 is driven by an input signal whose autocorrelation function is rectangular in shape—10 units high and extending from $\tau=-0.15\text{s}$ to $\tau=0.15\text{s}$. Find the cross-correlation function relating output with input, i.e. $R_{uy}(\tau)$ and sketch.

12 Find $R_{y_1y_2}(\tau)$ in terms of $g_1(t)$, $g_2(t)$ and $R_{x1x2}(\tau)$ for the systems shown.

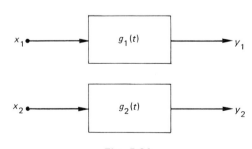

Fig. 5.34

13 The system described in question 10 is driven by an input signal whose autocorrelation function is of the form

$$R_{uu}(\tau) = 10ae^{-a|\tau|}.$$

Show that the input-output cross-correlation is given by

$$R_{uy}(\tau) = 5a\left[e^{-a\tau}\left(\frac{1}{2-a} - \frac{1}{4-a}\right) + e^{-2\tau}\left(\frac{1}{2+a} - \frac{1}{2-a}\right) - e^{-4\tau}\left(\frac{1}{4+a} - \frac{1}{4-a}\right)\right]$$

for $\tau \geqslant 0$

$$= 5ae^{a\tau}\left(\frac{1}{2+a} - \frac{1}{4+a}\right) \qquad \text{for } \tau \leqslant 0.$$

Hence demonstrate that as $a \to \infty$, i.e. $R_{uu}(\tau)$ tends to become an impulse, $R_{uy}(\tau)$ tends to become proportional to the impulse response of the system. Take $a = 20$ and reduce $R_{uy}(\tau)$ to a numerical function and sketch.

14 Find the output autocorrelation function $R_{yy}(\tau)$ for the systems and signals of (a) question 10 and (b) question 13 for $a = 20$. Sketch $R_{yy}(\tau)$ in each case.

6

Signals in Linear Systems—Frequency Domain

6.1   Introduction

We saw in Chapter 4 that arbitrary signals, and even random signals, can be regarded as either being made up of sinusoids or as having a sinusoidal content. Thus, when such signals form the input to a linear system it is natural to want to find how the system responds and how the various frequency components are handled to form the output. To investigate we find how the system responds to sinusoids of different frequencies. Thus we study the frequency response of a system and we find in general, that the amplitude change between input and output and the phase change depend on frequency. This dependence is the principal part of our study.

6.2   Frequency Response for Continuous Signals and Systems

The steady state output of a system can be determined from the convolution integral:

$$y(t) = \int_0^\infty g(\eta) \cdot u(t-\eta) \, d\eta.$$

If the input, u, is a sinusoid of unit amplitude applied in the infinite past then we can say

$$u(t) = e^{j\omega t} = \cos \omega t + j \sin \omega t$$

where ω is its angular frequency.

If, however, the input is a cosine signal then $u(t) =$ real part of $e^{j\omega t}$ and if it is a sine signal it will be the Imaginary Part of $e^{j\omega t}$.

Then

$$g(t) = \int_0^\infty g(\eta) e^{j\omega(t-\eta)} \, d\eta$$

$$= e^{j\omega t} \int_0^\infty g(\eta) e^{-j\omega \eta} \, d\eta. \tag{6.1}$$

However, $\int_0^\infty g(\eta) e^{-j\omega \eta} \, d\eta$ is the defining integral for the LT of $g(\eta)$ for the special case $s = j\omega$.

Therefore

$$y(t) = e^{j\omega t} G(j\omega). \tag{6.2}$$

Since $G(j\omega)$ is a complex number we can write it as $G(j\omega) = |G(j\omega)| \, \underline{/G(j\omega)}$ and the ratio of the output sinusoid to the input sinusoid will be

$$\frac{e^{j\omega t} G(j\omega)}{e^{j\omega t}} = G(j\omega). \tag{6.3}$$

We conclude that the ratio of the output amplitude to the input amplitude is $|G(j\omega)|$ and that the phase change is $\underline{/G(j\omega)}$, $G(j\omega)$ is sometimes known as the frequency transfer function or, if there is no ambiguity with $G(s)$, simply as the transfer function.

To clarify the meaning of equation (6.2) let us take the case when the input is a cosine wave. Then the corresponding output will be the real part of $e^{j\omega t}G(j\omega)$. Let $G(j\omega)=A+jB$ then the output will be the real part of:

$$(\cos\omega t+j\sin\omega t)(A+jB)=A\cos\omega t-B\sin\omega t.$$

Let
$$A\cos\omega t-B\sin\omega t=R\cos(\omega t+\phi)$$

$$=R\cos\omega t\cos\phi-R\sin\omega t\sin\phi.$$

Then $R\cos\phi=A$ and $R\sin\phi=B$ from which

$$R^2(\cos^2\phi+\sin^2\phi)=A^2+B^2$$

giving
$$R=\sqrt{A^2+B^2}=|G(j\omega)|$$

Also
$$\tan\phi=\frac{\sin\phi}{\cos\phi}=\frac{B}{A}$$

and
$$\phi=\arctan\frac{B}{A}=\underline{/G(j\omega)}.$$

Thus the output is

$$|G(j\omega)|\cos(\omega t+\underline{/G(j\omega)})$$

which confirms that the statement above is relevant to amplitude ratio and phase change.

A given system for which $G(j\omega)$ is known can easily have its frequency response evaluated by substituting different values of ω into $|G(j\omega)|$ and $\underline{/G(j\omega)}$ resulting in a series of pairs; one pair for each ω. The best way of displaying this information is graphically and there are a number of ways of doing this, the two most common being

(1) rectangular axes and
(2) polar axes.

In the first case we plot $20\log_{10}|G(j\omega)|$* versus ω using log–linear axes with ω horizontally on the log scale as in Fig. 6.1 and $\underline{/G(j\omega)}$ versus ω similarly. This pair of graphs is known as the Bode diagram after Henrich Bode who first drew attention to their value in frequency response manipulations.

In the second case $|G(j\omega)|$ is taken as radius and $\underline{/G(j\omega)}$ is taken as angle on polar axes as in Fig. 6.2. However, we will restrict our attention to the first case of rectangular axes.

6.3 Frequency Response from the Pole-Zero Map, Continuous Signals

In section 5.11 we found that we could evaluate $|G(s)|$ and $\underline{/G(s)}$ for varying values of s using the pole-zero map so obviously if we choose $s=j\omega_1$ (ω_1 being some value of ω) we could evaluate $|G(j\omega_1)|$ and $\underline{/G(j\omega_1)}$. Now the value $s=j\omega_1$ lies on the imaginary axis of the s-plane, as in Fig. 6.3, and it follows that all values for ω will form a locus coincident with this axis. It is easy, then, to see how to obtain data for a frequency response for a given TF.

* $20\log_{10}|G(j\omega)|$ is usually quoted in deciBels (dB).

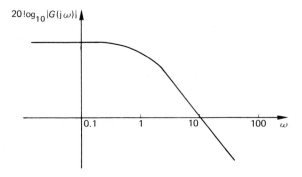

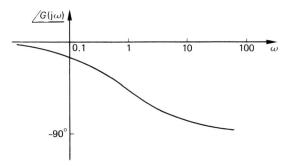

Fig. 6.1

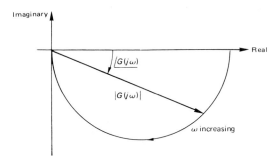

Fig. 6.2

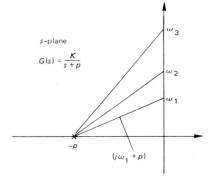

Fig. 6.3

For the example illustrated in Fig. 6.3 we see immediately that, as ω increases $|j\omega + p|$ increases causing $|K/j\omega + p|$ to decrease and when $\omega = \infty$ it is zero. Conversely, when $\omega = 0$, $|K/j\omega + p| = K/p$. Also as ω increases $\underline{/j\omega + p}$ increases (positively) causing $\underline{/K/j\omega + p}$ to increase negatively, when $\omega = \infty$, $\underline{/K/j\omega + p} = -90°$ and when $\omega = 0$, $\underline{/K/j\omega + p} = 0$. The corresponding Bode diagram for this simple first order TF is that shown in Fig. 6.1.

Clearly this idea can be extended to more complicated TFs but since Bode diagrams can be constructed in another way, the above method would not normally be used. However, the value of the s-plane method is to assess the behaviour of a system in frequency response for different pole-zero locations.

A further example, $G(s) = \dfrac{5}{s^2 + 2s + 5}$ from which we notice that it is the prototype second order system with $\omega_n = \sqrt{5}$ and $\zeta = 0.447$.

Then
$$G(j\omega) = \frac{5}{(j\omega)^2 + 2j\omega + 5} = \frac{6}{(j\omega + 1 + j2)(j\omega + 1 - j2)}$$

and its pole-zero map is shown in Fig. 6.4.

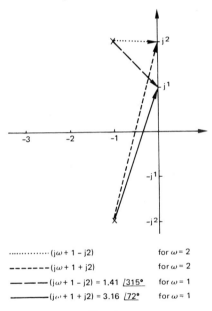

$\cdots\cdots\cdots\cdots (j\omega + 1 - j2)$	for $\omega = 2$
$------ (j\omega + 1 + j2)$	for $\omega = 2$
$--- (j\omega + 1 - j2) = 1.41 \,\underline{/315°}$	for $\omega = 1$
$---(j\omega + 1 + j2) = 3.16 \,\underline{/72°}$	for $\omega = 1$

Fig. 6.4

Typically
$$|G(j1)| \simeq \frac{5}{3.16 \times 1.41} = 1.12$$

$$\underline{/G(j1)} \simeq -72° - 315° = -27°.$$

Note that for $\omega = 2$, $(j\omega + 1 - j2)$ is a minimum causing $|G(j2)|$ to be near to maximum, thus $\omega \simeq 2$ is the resonant frequency for this system. It is easily deduced that

when $\omega = 0$,

$$|G(j\omega)| = \frac{5}{\sqrt{5} \times \sqrt{5}} = 1 \qquad \text{and} \qquad \underline{/G(j\omega)} = -63.4 - 296.6 = 0°$$

when $\omega = \infty$

$$|G(j\omega)| = \frac{5}{\infty \times \infty} = 0 \quad \text{and} \quad \underline{/G(j\infty)} = -90 - 90 = -180°.$$

For this example we can write

$$G(j\omega) = \frac{5}{(5 - \omega^2) + 2j\omega} \quad \text{since } (j\omega)^2 = -\omega^2$$

$$|G(j\omega)| = \frac{5}{[(5 - \omega^2)^2 + 4\omega^2]^{1/2}}$$

$$\underline{/G(j\omega)} = -\arctan\frac{2\omega}{5 - \omega^2}$$

and in general for the prototype second order system

$$G(s) = \frac{\omega_n^2}{s^2 + 2\zeta\omega_n s + \omega_n^2}$$

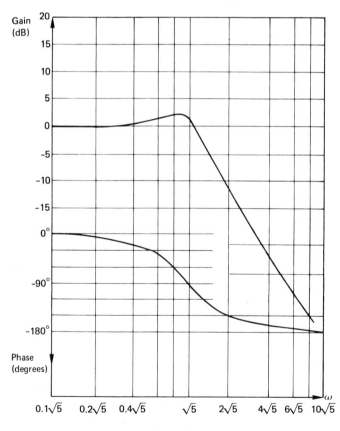

Fig. 6.5

or
$$G(j\omega) = \frac{\omega_n^2}{(j\omega)^2 + 2\zeta\omega_n j\omega + \omega_n^2}$$

$$G(j\omega) = \frac{\omega_n^2}{(\omega_n^2 - \omega^2) + 2\zeta\omega_n j\omega}$$

$$|G(j\omega)| = \frac{\omega_n^2}{[(\omega_n^2 - \omega^2)^2 + 4\zeta^2\omega_n^2\omega^2]^{1/2}}$$

$$\underline{/G(j\omega)} = -\arctan\frac{2\zeta\omega_n\omega}{\omega_n^2 - \omega^2}.$$

The Bode diagram for the numerical example is shown in Fig. 6.5.

6.4   Convolution and the Fourier Transform

The work in section 6.2 and the form of equation (6.1) recalls the Fourier Transform (FT) of Chapter 4 which naturally prompts us to look at the FT in this context. We start by looking at it in the context of convolution as we did for the LT and the ZT.

As usual we have the output of a system given by the convolution integral

$$y(t) = \int_0^\infty g(\eta)\, u(t - \eta)\, d\eta$$

for the steady state condition after the input was applied in the infinite past. Now take the Fourier Transform FT

$$\mathscr{F}[y(t)] = \int_{-\infty}^\infty \int_0^\infty g(\eta)\, u(t - \eta)\, d\eta\, e^{-j\omega t}\, dt.$$

Multiplying the integrand by $e^{-j\omega\eta}\, e^{j\omega\eta}$ we get

$$\mathscr{F}[y(t)] = \int_{-\infty}^\infty \int_0^\infty g(\eta)\, e^{-j\omega\eta}\, u(t - \eta)\, e^{-j\omega(t - \eta)}\, d\eta\, dt.$$

Now define $\lambda = t - \eta$ and get

$$\mathscr{F}[y(t)] = \int_{-\infty}^\infty \int_0^\infty g(\eta)\, e^{-j\omega\eta}\, u(\lambda)\, e^{-j\omega\lambda}\, d\eta\, d\lambda.$$

Since we are concerned only with realisable systems $g(t) = 0$ for $t < 0$, and keeping this always in mind, allows us to replace the lower limit of zero in the inner integral by $-\infty$ to obtain

$$\mathscr{F}[y(t)] = \int_{-\infty}^\infty g(\eta)\, e^{-j\omega\eta}\, d\eta \int_{-\infty}^\infty u(\lambda)\, e^{-j\omega\lambda}\, d\lambda$$

which we recognise as the product of two FTs, thus

$$\mathscr{F}[y(t)] = Y(j\omega) = G(j\omega)\, U(j\omega).$$

Whilst we have developed this for steady state conditions (because our principal use of it will be in that context) we could show, by using the step function as we did for LT and ZT, that it is valid for any condition. It is also valid for any signal, not just a sinusoid. If, however, u is a cosine applied in the infinite past then $u = \cos\omega_0 t$ and

$U(j\omega) = \pi(\delta(\omega - \omega_0) + \delta(\omega + \omega_0))$ giving
$$Y(j\omega) = G(j\omega)\pi[\delta(\omega - \omega_0) + \delta(\omega + \omega_0)]$$

which tells us that y is a sinusoid with amplitude $|G(j\omega)|$ and phase shift $\underline{/G(j\omega)}$ confirming our earlier result.

This demonstrates the value of the FT for systems with sinusoidal inputs in steady state conditions and is a more concise result than that obtained using the LT.

6.5   Frequency Response for Discrete Signals and Systems

We can see the same principles here as we did for continuous signals and systems. Recall equation (5.11a), the discrete convolution sum:

$$y[k] = \sum_{i=0}^{k} g[i]\, u[k-i].$$

The steady state case is when the input began in the infinite past causing $k = \infty$ at the upper limit, then

$$y[k] = \sum_{i=0}^{\infty} g[i]\, u[k-i].$$

An input sinusoid (discrete) can be written as $e^{ji\omega\Delta}$ giving

$$y[k] = \sum_{i=0}^{\infty} g[i]\, e^{j(k-i)\omega\Delta}$$

$$= e^{jk\omega\Delta} \sum_{i=0}^{\infty} g[i]\, e^{-ji\omega\Delta}$$

$$= e^{jk\omega\Delta} \sum_{i=0}^{\infty} g[i]\, z^{-i} \qquad \text{for } z = e^{j\omega\Delta}.$$

By definition

$$\sum_{i=0}^{\infty} g[i]\, z^{-i} \quad \text{is } G(z) \qquad \text{for } z = e^{j\omega\Delta}$$

then $\qquad\qquad\qquad y[k] = e^{jk\omega\Delta}\, G(z) \qquad \text{for } z = e^{j\omega\Delta}.$

The same principles hold for cosine and sine inputs regarding real and imaginary components and, of course, $G(z) = G(e^{j\omega\Delta})$ is complex with modulus $|G(e^{j\omega\Delta})|$ and arguments $\underline{/G(e^{j\omega\Delta})}$ which can be evaluated for different values of ω and plotted as before.

6.6   Frequency Response from the Pole-Zero Map, Discrete Signals

We know, of course, that $|G(z)|$ and $\underline{/G(z)}$ can be evaluated using the pole-zero map in the z-plane for different values of z. In our present case z takes on values determined by $z = e^{j\omega\Delta}$ but we know that $|e^{j\omega\Delta}| = 1$ and that $e^{j\omega\Delta} = \cos\omega\Delta + j\sin\omega\Delta$ giving $\underline{/e^{j\omega\Delta}} = \omega\Delta$, Fig. 6.6. It follows that, as ω increases from zero to infinity, z (on the z-plane) traces out a circle of unit radius but when ω is an integral multiple of 2π we have $e^{j0} = e^{j2\pi} = e^{j4\pi} = 1$, whereas on the s-plane each value of ω identifies a unique position for s. We find that on the z-plane z appears at the same point on the unit circle for ω, $\omega + \dfrac{2\pi}{\Delta}$, $\omega + \dfrac{4\pi}{\Delta}$, $\dfrac{\omega + 2n\pi}{\Delta}$ (where

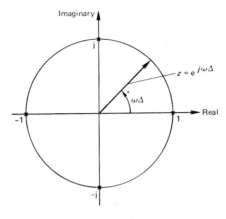

Fig. 6.6

n is an integer), giving rise to ambiguity which demands care in the use of frequency response data for discrete signals. This ambiguity is highlighted when we use the z-plane pole-zero map for evaluation. Consider a simple first order system $G(z)=\dfrac{Kz}{z-\phi}$ mapped in Fig. 6.7.

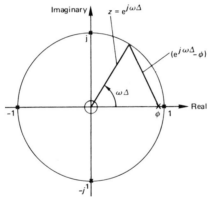

Fig. 6.7

Then

$$|G(\mathrm{e}^{\mathrm{j}\omega\Delta})|=\frac{K\times 1}{|\mathrm{e}^{\mathrm{j}\omega\Delta}-\phi|}$$

and

$$\underline{/G(\mathrm{e}^{\mathrm{j}\omega\Delta})}=\omega\Delta-\underline{/\mathrm{e}^{\mathrm{j}\omega\Delta}-\phi}$$

However, when we have $\omega+\dfrac{2\pi}{\Delta}$, $\omega+\dfrac{4\pi}{\Delta}$, $\omega+\dfrac{2n\pi}{\Delta}$, we get the same values for $|G(\mathrm{e}^{\mathrm{j}\omega\Delta})|$ and $\underline{/G(\mathrm{e}^{\mathrm{j}\omega\Delta})}$.

This causes the graphical representation to show a periodicity in the frequency response data as illustrated in Fig. 6.8.

This means that, for normal applications, we must restrict our range of frequencies to the range 0 to π/Δ meaning that Δ is a very important parameter in discrete system design.

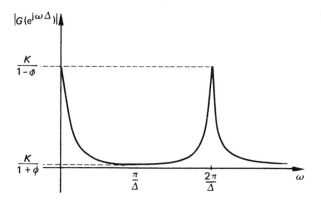

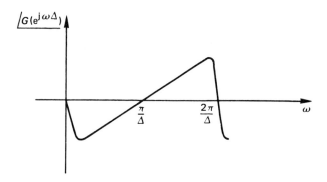

Fig. 6.8

(Nyquist's sampling theory says that the sampling frequency for a signal must be at least twice the highest frequency component in the signal. The sampling frequency is $2\pi/\Delta$, thus the maximum frequency component of the input to the above system is $\dfrac{\pi}{\Delta}$ which tallies with the restriction mentioned. This theorem is also due to Shannon).

This problem is illustrated in Fig. 6.9 where we show the output of a system which could be obtained by either of two inputs. (Of course, even higher frequencies could give the same output). This effect is known as aliasing and often digital systems have anti-aliasing filters to remove frequencies above the limiting frequency to avoid ambiguous results. As usual we

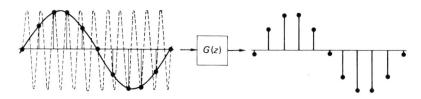

Fig. 6.9

illustrate with an example. Let $G(z) = \dfrac{10z}{z^2 - 0.4z + 0.4}$ (a prototype second order system with

$\mu = 0.632$ and $\gamma = 0.316$ where $\mu = e^{-p\Delta}$ and $\gamma = \cos \omega\Delta$.

This can be written as

$$G(z) = \frac{10z}{(z - 0.2 + \text{j}0.6)(z - 0.2 - \text{j}0.6)}$$

for which the pole-zero map is as shown in Fig. 6.10 and the system is obviously stable. If $\Delta = 0.2$ then for $\omega = 1$ we get $\omega\Delta = 0.2$ or $11.5°$ approximately and

$$|G(e^{\text{j}1 \times 0.2})| \simeq \frac{10}{1.1 \times 0.85}$$

$$= 10.695$$

$$\underline{/G(e^{\text{j}1 \times 0.2})} \simeq 11.5° - 333° - 48°$$

$$= -369.5°$$

$$\therefore \underline{/G(e^{\text{j}1 \times 0.2})} = -9.5°.$$

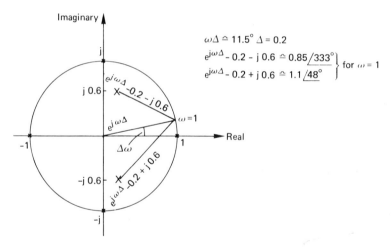

Fig. 6.10

Notice that for $\omega\Delta = 1.249 = (71.565°)$, that is $\omega = 6.245$, $|e^{\text{j}\omega\Delta} - 0.2 - \text{j}0.6|$ is a minimum causing $|G(e^{\text{j}\omega\Delta})|$ to be approximately at maximum. Thus the frequency response shows band pass characteristics as did the equivalent continuous system. Remember that for frequencies greater than $\dfrac{2\pi}{\Delta}$, that is for $\omega \to \dfrac{2\pi}{\Delta} = \dfrac{2\pi}{0.2} = 31.4 \text{ rad s}^{-1}$, the response repeats becoming periodic in this frequency and that the upper limit of usable frequency is $\dfrac{\pi}{\Delta} \simeq 15.7 \text{ rad s}^{-1}$. The corresponding Bode plot is shown in Fig. 6.11.

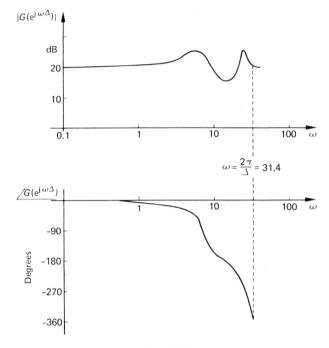

Fig. 6.11

6.7   Input–Output Cross–Correlation in the Frequency Domain

In section 5.13 we found that the input–output cross–correlation was given by equation (5.22)

$$R_{uy}(\tau) = \int_0^\infty g(\eta)R_{uu}(\tau - \eta)d\eta$$

and noted that it was in the form of a convolution integral. In Chapter 4 we found that cross-spectral density and cross-correlation were a FT pair, thus on taking the FT of $R_{uy}(\tau)$ we will get the input–ouput cross-spectral density as follows:

$$\mathscr{F}[R_{uy}(\tau)] = S_{uy}(\omega) = \int_{-\infty}^\infty \int_{-\infty}^\infty g(\eta)R_{uu}(\tau - \eta)d\eta e^{j\omega\tau}d\tau.$$

On using the routine developed in section 6.4 we will get

$$S_{uy}(\omega) = G(j\omega)S_{uu}(\omega).$$

This tells us that we get $S_{uy}(\omega)$ by multiplying the frequency TF by the spectral density of the input.

Example

For comparison we will use the same arrangement as in Fig. 5.26 in which

$$R_{uu}(\lambda) = 4e^{-3|\lambda|} \qquad \text{and} \qquad g(\eta) = 2e^{-2\eta}$$

Using a table of FTs we get

$$S_{uu}(\omega) = \frac{24}{9+\omega^2} \quad \text{and} \quad G(j\omega) = \frac{2}{2+j\omega}$$

giving

$$S_{uy}(\omega) = \frac{48}{(9+\omega^2)(2+j\omega)}$$

Polar and rectangular plots are shown in Fig. 6.12.

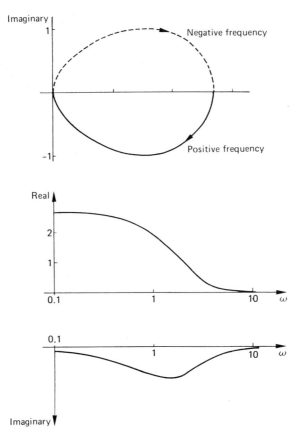

Fig. 6.12

If we need to find $R_{uy}(\tau)$ it is necessary to perform the inverse FT of this as follows:
Using partial fractions we get

$$S_{uy}(\omega) = \frac{48/5}{j\omega+2} - \frac{8}{j\omega+3} + \frac{8/5}{-j\omega+3}$$

From the properties of FT we note that the first two terms on the r.h.s. involve $+\omega$ and are relevant to positive time whilst the last term involves $-\omega$ and is relevant to negative time so that inverse transformation must be taken in two parts. This gives

$$R_{uy}(\tau) = 9\tfrac{3}{5}e^{-2\tau} - 8e^{-3\tau} \qquad \text{for } \tau \geqslant 0$$

$$= 1\tfrac{3}{5}e^{3\tau} \qquad \text{for } \tau \leqslant 0$$

and is the same result as we obtained at equations (5.23) and (5.24) using the time domain solution method.

Even including the rather tedious partial fraction working (not shown above) this is a much more convenient method of solution than previously.

For the special case where the input is white noise with $S_{xx}(\omega) = N$ we get, (where N is a constant whose dimensions are those of spectral density, (signal quantity^2 × time).

$$S_{uy}(\omega) = G(j\omega) . N$$

and the corresponding conclusion to that obtained in section 5.13 is that if a system is excited by white noise then the input–output cross-spectral density is proportional to the frequency TF. This is the frequency domain equivalent of system identification by cross-correlation, namely

$$G(j\omega) = \frac{S_{uy}(\omega)}{N}$$

which means that $G(j\omega)$ can be obtained by observing $S_{uy}(\omega)$ with a white noise input.

To work this problem using discrete signals requires the use of the DFT (or FFT) which is, of course, too involved for manual calculation; however, the approach would be the same as that outlined in the solution to the continuous problem. Clearly, if $S_{uu}(\omega)$ and $G(j\omega)$ were known at the outset in discrete form then evaluation of $S_{uy}(\omega)$ would be easy and no transformations would be needed provided that only $S_{uy}(\omega)$ was required and not $R_{uy}(\tau)$.

6.8   Output Auto-correlation in the Frequency Domain

In section 5.14 at equation (5.30) we had

$$R_{yy}(\tau) = \int_0^\infty \int_0^\infty g(\eta)g(\lambda)R_{uu}(\tau - \lambda + \eta)\mathrm{d}\lambda\mathrm{d}\eta$$

but here we want to find the frequency domain equivalent. As in the previous section we will take the FT:

$$\mathscr{F}[R_{yy}(\tau)] = S_{yy}(\omega) = \int_{-\infty}^\infty \int_{-\infty}^\infty \int_{-\infty}^\infty g(\eta)g(\lambda)R_{uu}(\tau - \lambda + \eta)\mathrm{d}\lambda\mathrm{d}\eta\,e^{-j\omega\tau}\mathrm{d}\tau.$$

On multiplying the integrand by $e^{j\omega\eta}e^{-j\omega\eta}e^{j\omega\lambda}e^{-j\omega\lambda}$ we can put

$$S_{yy}(\omega) = \int_{-\infty}^\infty \int_{-\infty}^\infty \int_{-\infty}^\infty g(\eta)e^{j\omega\eta}g(\lambda)e^{-j\omega\lambda}R_{uu}(\tau - \lambda + \eta)e^{-j\omega(\tau - \lambda + \eta)}\mathrm{d}\lambda\mathrm{d}\eta\mathrm{d}\tau.$$

On making the substitution $\gamma = \tau - \lambda + \eta$ we get

$$S_{yy}(\omega) = \int_{-\infty}^\infty g(\eta)e^{j\omega\eta}\mathrm{d}\eta \int_{-\infty}^\infty g(\lambda)e^{-j\omega\lambda}\mathrm{d}\lambda \int_{-\infty}^\infty R_{uu}(\gamma)e^{-j\omega\gamma}\mathrm{d}\gamma.$$

The first integral we recognise as the definition of $G(-j\omega)$ the second that of $G(j\omega)$ and the third that of $S_{uu}(\omega)$

Therefore

$$S_{yy}(\omega) = G(-j\omega)G(j\omega)S_{uu}(\omega)$$

$$= |G(j\omega)|^2 S_{uu}(\omega).$$

In our numerical example we have

$$S_{uu}(\omega) = \frac{24}{9+\omega^2} \quad \text{and} \quad |G(j\omega)|^2 = \frac{4}{4+\omega^2}$$

Therefore

$$S_{yy}(\omega) = \frac{96}{(9+\omega^2)(4+\omega^2)}$$

and using partial fractions to find $R_{yy}(\tau)$ we get

$$S_{yy}(\omega) = \frac{96/5}{4+\omega^2} - \frac{96/5}{9+\omega^2}.$$

Referring to a table of transforms we find

$$\frac{96/5}{4+\omega^2} \rightleftharpoons \frac{96}{20}e^{-2|\tau|}$$

$$\frac{96/5}{9+\omega^2} \rightleftharpoons \frac{96}{30}e^{-3|\tau|}$$

$$R_{yy}(\tau) = \frac{24}{5}e^{-2|\tau|} - \frac{16}{5}e^{-3|\tau|}$$

which tallies with the result obtained using time domain methods and, once again, we see that the frequency domain method is far easier for calculation purposes. Fig. 6.13 shows $S_{yy}(\omega)$.

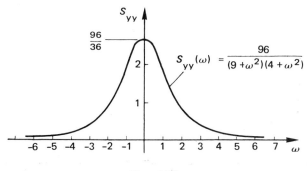

Fig. 6.13

6.9   Comment Relevant to Discrete Signals

Obviously the material of the last two sections could be re-considered for discrete signals using DFT but since this is best illustrated using computers, no attempt at illustrating the FFT by manual computation will be made. The principles are, of course, the same but the importance of sampling interval in relation to the frequencies involved must not be overlooked.

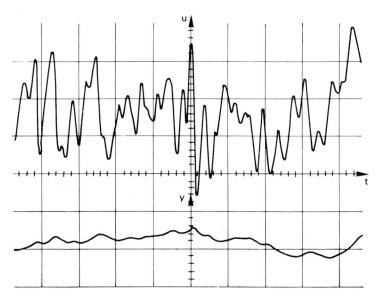

Fig. 6.14

Fig. 6.15

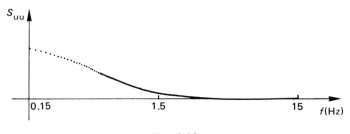

Fig. 6.16

Figs. 6.14–6.22 illustrate the corresponding results obtained on a digital signal analyser for discrete signals using the functions of the previous examples. The numerical examples worked in sections 5.13, 5.14, 6.7 and 6.8 are confirmed by experimental results illustrated in Figs. 6.14 to 6.22. These were obtained by setting up a simulation of the first order system whose impulse response is $2e^{-2n}$ and driving it with gaussian noise whose auto-correlation

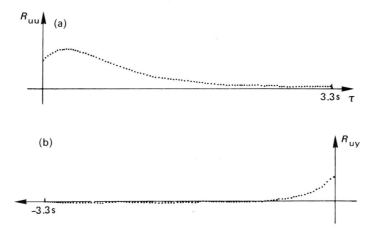

Fig. 6.17

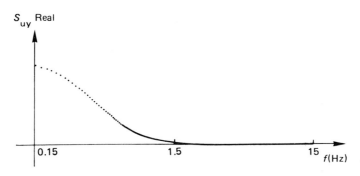

Fig. 6.18

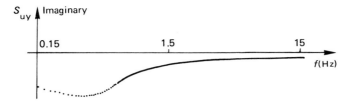

Fig. 6.19

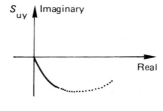

Fig. 6.20

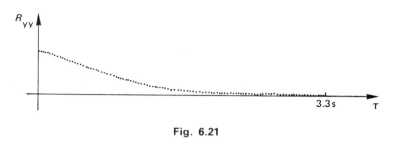

Fig. 6.21

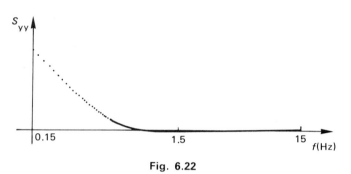

Fig. 6.22

function was $4e^{-3|\lambda|}$ as in Fig. 5.26 and processing the signals on a correlator and spectrum display unit in the manner described in section 9.2 and Fig. 9.3.

Fig. 6.14 shows the input (upper trace) and output (lower) signals unprocessed. Fig. 6.15 shows the positive time half of R_{uu} and is clearly exponential whilst its Fourier transform S_{uu} is shown in Fig. 6.16 with a logarithmic frequency scale. Fig. 6.17 shows R_{uy} for both positive and negative time and shows the same characteristic shape as expected from the calculations. Figs. 6.18 and 6.19 show the real and imaginary parts of S_{uy} respectively and, finally, Fig. 6.20 shows S_{uy} on polar axes, all corresponding with the calculated results. Fig. 6.21 shows the positive time half of R_{yy} and its corresponding S_{yy} is in Fig. 6.22.

We have not attempted to scale these results but the principles are discussed in section 9.2; however, the characteristic shapes are as expected.

6.10   Windows

If we are studying a continuous or discrete signal, which extends over a long time period it may be necessary for practical analysis to consider only a certain length and then regard this as an adequate representation of the whole. For example, for theoretical studies we can discuss signals of infinite duration and arrive at certain results; however, for practical measurements we can only take a finite length. The choice of the length of the measurement period is the subject of section 7.4.4 and does not affect us at this stage but it is appropriate to discuss here certain effects this truncation has on measurements.

If we refer to Fig. 6.23 we see that the selection of a certain length of signal from a much longer one can be regarded as the multiplication of the original long signal by another signal (or function) which has a much shorter duration. The latter is known as a window (or window function) and it is obvious why it is so called.

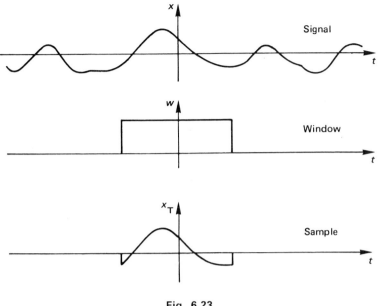

Fig. 6.23

Thus if the signal x is represented by $x(t)$, the window by $w(t)$, then the sample signal is given by

$$x_T(t) = x(t) \cdot w(t)$$

where the suffix on $x_T(t)$ is intended to convey the idea that it is a signal of duration T.

As one would expect, the properties of the signal x_T are different from those of x and it is therefore important to be aware of this and make due allowance when drawing conclusions about x from those obtained from x_T. In our deliberations we will find that the 'rectangular' window illustrated in Fig. 6.23 is not necessarily the best 'shape' for a given analysis; consequently, other windows are used and have been given names, emanating either from their shape or the person who first suggested it. Whilst Fig. 6.24 illustrates a few commonly used windows we will only examine the general properties and illustrate using the rectangular window.

An obvious question that immediately arises is: what effect does windowing have on the frequency spectrum of a signal? A start in investigating this can be made by comparing the FT of x with that of x_T, thus we will examine the latter.

The inverse Fourier Transform is

$$x(t) = \frac{1}{2\pi} \int_{-\infty}^{\infty} X(j\omega) e^{j\omega t} \, d\omega$$

and

$$w(t) = \frac{1}{2\pi} \int_{-\infty}^{\infty} W(j\omega) e^{j\omega t} \, d\omega.$$

We can then put

$$x(t)w(t) = \frac{1}{4\pi^2} \int_{-\infty}^{\infty} X(j\omega) e^{j\omega t} \, d\omega \int_{-\infty}^{\infty} W(j\Omega) e^{j\Omega t} \, d\Omega.$$

(we need to introduce Ω to keep tabs on the variables of integration, a concept we have already met).
Then

$$x(t) \cdot w(t) = \frac{1}{4\pi^2} \int_{-\infty}^{\infty} \int_{-\infty}^{\infty} X(j\omega) W(j\Omega) e^{j\omega t} e^{j\Omega t} \, d\omega \, d\Omega$$

if we write $\Omega = v - \omega$ (a method we have used before) then we get

$$x(t) \cdot w(t) = \frac{1}{4\pi^2} \int_{-\infty}^{\infty} \int_{-\infty}^{\infty} X(j\omega) W(jv - j\omega) e^{j\omega t} e^{j(v - \omega)t} \, d\omega \, dv$$

$$= \frac{1}{4\pi^2} \int_{-\infty}^{\infty} \int_{-\infty}^{\infty} X(j\omega) W(jv - j\omega) e^{jvt} \, d\omega \, dv$$

$$= \frac{1}{4\pi^2} \int_{-\infty}^{\infty} \left[\int_{-\infty}^{\infty} X(j\omega) W(jv - j\omega) \, d\omega \right] e^{jvt} \, dv.$$

If we replace the integral in brackets by $X'_T(jv)$ we get

$$x(t) \cdot w(t) = \frac{1}{2\pi} \left\{ \frac{1}{2\pi} \int_{-\infty}^{\infty} X'_T(jv) e^{jvt} \, dv \right\}.$$

We recognise the integral in braces as the inverse FT of some signal $x'_T(t)$ giving

$$x(t) \cdot w(t) = \frac{1}{2\pi} x'_T(t) = x_T(t)$$

from which we conclude that

$$\mathscr{F}[x(t) \cdot w(t)] = \mathscr{F}[x_T(t)] = X_T(j\omega) = \frac{1}{2\pi} \int_{-\infty}^{\infty} X(j\omega) W(jv - j\omega) \, d\omega$$

and that the integral on the r.h.s. is a convolution integral. This means that the FT of the product of $x(t)$ and $w(t)$ is $\frac{1}{2\pi} \times$ the convolution of the FT of $x(t)$ and that of $w(t)$ and illustrates again the relationship between multiplication and convolution. However, in this instance, multiplication is in the time domain and convolution in the frequency domain which is the reverse of what we have found already. Fig. 6.24 shows the FT of the windows illustrated and the derivation of that for the rectangular window follows that developed in section 4.4.

We now illustrate by means of an example. Our signal x will be an infinite cosine wave and our window will be the simplest, namely the rectangular. Fig. 6.25 illustrates this.

Whilst we have shown the duration of the window, ($-T$ to $+T$), corresponding to one period of the cosine we will work the example for any duration and then see the effect of relating it to the relevant period.

Referring to a table of transforms we find

$$\mathscr{F}(\cos \omega_0 t) = \pi[\delta(\omega - \omega_0) + \delta(\omega + \omega_0)]$$

$$\mathscr{F}[\text{Rect. Window}] = 2T \frac{\sin \omega T}{\omega T}$$

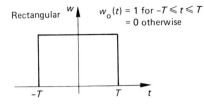

Rectangular $w_0(t) = 1$ for $-T \leqslant t \leqslant T$
$= 0$ otherwise

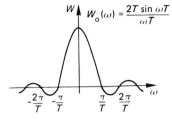

$W_0(\omega) = \dfrac{2T \sin \omega T}{\omega T}$

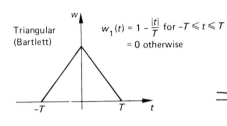

Triangular (Bartlett) $w_1(t) = 1 - \dfrac{|t|}{T}$ for $-T \leqslant t \leqslant T$
$= 0$ otherwise

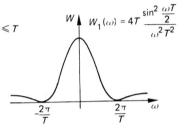

$W_1(\omega) = 4T \dfrac{\sin^2 \dfrac{\omega T}{2}}{\omega^2 T^2}$

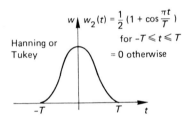

$w_2(t) = \dfrac{1}{2}\left(1 + \cos \dfrac{\pi t}{T}\right)$
for $-T \leqslant t \leqslant T$
$= 0$ otherwise

Hanning or Tukey

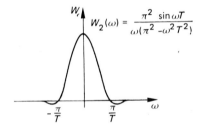

$W_2(\omega) = \dfrac{\pi^2 \sin \omega T}{\omega(\pi^2 - \omega^2 T^2)}$

$W_3(\omega) = \dfrac{(1.08\pi^2 - 0.16T^2 \pi^2)\sin \omega T}{\omega(\pi^2 - \omega^2 T^2)}$

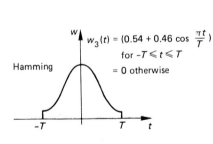

$w_3(t) = (0.54 + 0.46 \cos \dfrac{\pi t}{T})$
for $-T \leqslant t \leqslant T$
$= 0$ otherwise

Hamming

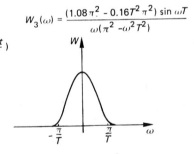

Fig. 6.24

and using the convolution integral we get

$$\mathscr{F}[\cos \omega t \times (\text{Rect. Window})] = \int_{-\infty}^{\infty} [\delta(\omega - \omega_0) + \delta(\omega + \omega_0)] T \frac{\sin(v - \omega)T}{(v - \omega)T} d\omega.$$

Remembering the sifting properties of the impulse function (that is, in this case, $\left[\dfrac{[\sin(v - \omega)T]}{(v - \omega)T}\right]$ will only contribute to the integrand when $\omega = \omega_0$ and $\omega = -\omega_0$) we can

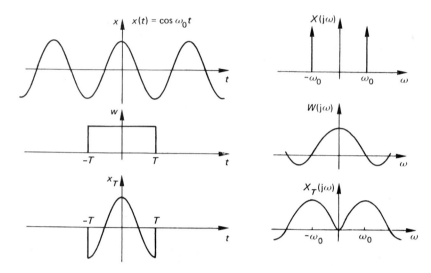

Fig. 6.25

simplify the integral to

$$\mathscr{F}(x_T) = T\frac{\sin(v-\omega_0)T}{(v-\omega_0)T}\int_{-\infty}^{\infty}\delta(\omega-\omega_0)d\omega$$

$$+ T\frac{\sin(v+\omega_0)T}{(v+\omega_0)T}\int_{-\infty}^{\infty}\delta(\omega+\omega_0)d\omega$$

$$= T\frac{\sin(\omega-\omega_0)T}{(\omega-\omega_0)T}+T\frac{\sin(\omega+\omega_0)T}{(\omega+\omega_0)T}.$$

(We have replaced the dummy variable v by ω).

Thus we conclude that the impulsive spectrum of $x(t)=\cos\omega t$ is replaced by the twin narrow band continuous spectrum of $x_T(t)$. Fig. 6.26 illustrates this for different values of T, from which we get the obvious conclusion that a better representation of $X(j\omega)$ is obtained by making T as large as possible. We see that the effect of the rectangular window is to introduce side lobes of alternating sign which are quite significant in size and which is clearly a disadvantage.

A motivation for inventing windows other than the rectangular one is to reduce the significance of these lobes and this effect is illustrated in Fig. 6.24 where the FT of each window is shown. The higher order windows have smaller lobes which produce smaller ones in the FT of the sample.

We have developed the principles using windows which are symmetrical about the $t=0$ axis because this makes the derivations simpler and this has meant the inclusion of negative frequencies, of course. In practice the window can only exist in positive time and the concept of negative frequency only appears in calculations but, nevertheless, the effect of windows in practice is the same as that illustrated and is, clearly, an effect that cannot be ignored.

Whilst we have developed the ideas for a signal x a little thought makes us realise that windows are also relevant to the relationships between auto-correlation and spectral density because they are related to the FT. Clearly, then, in obtaining the spectral density of a signal from its auto-correlation function, in practice we have to take the FT of a windowed

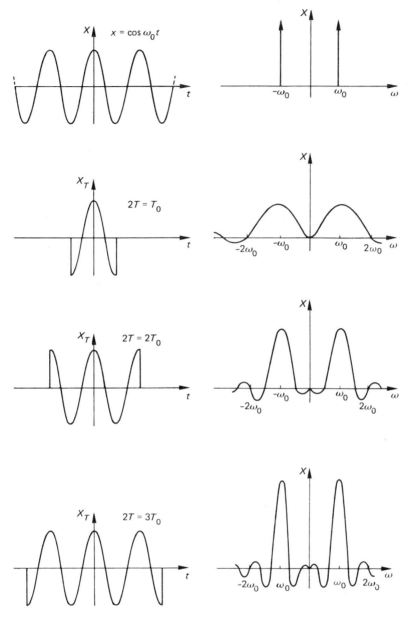

Fig. 6.26

version of the latter. This will result in side lobes in the spectral density and, if the duration and/or shape of the window is not carefully chosen, serious errors can arise. We will illustrate by means of an example.

If we take the complete exponential auto-correlation function $R_{xx}(\tau) = e^{-a|\tau|}$ we find (from a table of transforms) that its spectral density is $S_{xx}(\omega) = \dfrac{2a}{(a^2 + \omega^2)}$ (see Fig. 6.27).

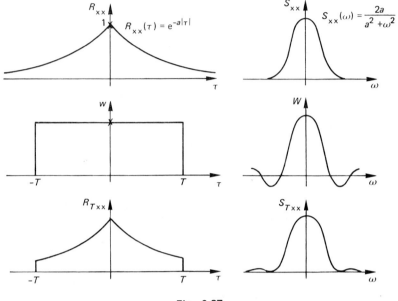

Fig. 6.27

To proceed in the manner of the previous example we would begin by writing

$$\mathscr{F}[R_{xx} \times (\text{Rect. Window})] = \mathscr{F}[R_{Txx}(\tau)] = \frac{1}{2\pi}\int_{-\infty}^{\infty} \frac{2a}{a^2+\omega^2} 2T\frac{\sin(v-\omega)T}{(v-\omega)T}d\omega.$$

Unlike the previous convolution integral this is not exactly easy to evaluate so we will resort to the practical device of evaluating $\mathscr{F}[R_{Txx}(\tau)]$ directly. Thus we get

$$\mathscr{F}[R_{Txx}(\tau)] = \int_{-\infty}^{\infty} e^{-a|\tau|} \times (\text{Rect. Window})e^{-j\omega\tau}d\tau.$$

Since the effect of the window is to make the integrand zero for $\tau < -T$ and $\tau > T$ we can re-write as

$$\mathscr{F}[R_{Txx}(\tau)] = \int_{-T}^{T} e^{-a|\tau|}e^{-j\omega\tau}d\tau$$

$$= \int_{-T}^{0} e^{a\tau}e^{-j\omega\tau}d\tau + \int_{0}^{T} e^{-a\tau}e^{-j\omega\tau}d\tau$$

$$= \int_{-T}^{0} e^{(a-j\omega)\tau}d\tau + \int_{0}^{T} e^{(-a-j\omega)\tau}d\tau$$

$$= \left[\frac{e^{(a-j\omega)\tau}}{a-j\omega}\right]_{-T}^{0} + \left[\frac{e^{(-a-j\omega)\tau}}{-a-j\omega}\right]_{0}^{T}$$

$$= \frac{1-e^{-(a-j\omega)T}}{a-j\omega} - \frac{e^{(-a-j\omega)T}-1}{a+j\omega}$$

$$= \frac{1-e^{-aT}(\cos\omega T + j\sin\omega T)}{a-j\omega} - \frac{e^{-aT}(\cos\omega T - j\sin\omega T)-1}{a+j\omega}.$$

After a little tedious but straightforward algebraic manipulation we find that this reduces
to

$$\mathscr{F}\left[R_{Txx}(\tau)\right]=\frac{2a}{a^2+\omega^2}-\frac{2e^{-aT}(a\cos\omega T-\omega\sin\omega T)}{a^2+\omega^2}.$$

We note that the first term is the true spectrum and that the second term is an error which is
a sinusoid of decreasing amplitude. Clearly, if T is made large the error term is small and in
the limit if T is infinite the error is zero.

It is obvious, but nevertheless requires stressing, that windows will also have an effect in
discrete calculations.

Sometimes it is necessary to pre-filter a signal before it is processed to remove unwanted
components. This is often called frequency windowing and if the frequency spectrum of the
original signal is required the effect of the filter will emerge in the result. Again this effect
must be taken into account. However, we have discussed at length the effect of linear
systems on a signal and a filter is such a system; thus we have already discussed many
aspects of frequency windowing.

However, one aspect that requires mentioning arises due to the way in which some signal
analysers calculate auto-correlation. The signal is received and a sample taken (digitally);
thus it is subject to a time window. The FFT algorithm is applied to obtain $S_{xx}(\omega)$ (see
Chapter 7) after which the inverse FFT of $S_{xx}(\omega)$ is produced, which necessarily requires a
limited range of frequencies being used; thus spectral density is windowed. There are two
sources of window error (1) the time window on the original signal and (2) the window on
the spectral density. We have discussed the former and discovered its effect, principally the
introduction of lobes into the frequency spectrum, but a few words on the latter is in order.

Take the spectral density $S_{xx}(\omega)$ and produce the sample $S_{\Omega xx}(\omega)$ by multiplying by a
rectangular frequency window $F(\omega)$. See Fig. 6.28. The corresponding time function for $F(\omega)$

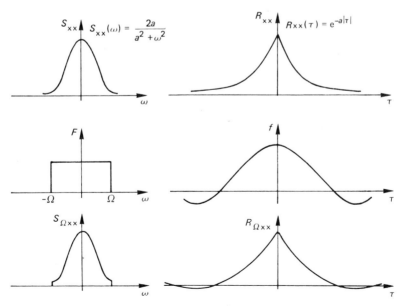

Fig. 6.28

is

$$f(\tau) = \mathscr{F}^{-1}[F(\omega)] = \frac{1}{2\pi} \int_{-\infty}^{\infty} F(\omega)e^{j\omega\tau}\,d\omega$$

which, for the rectangular window shown, is

$$f(\tau) = \frac{\Omega}{\pi} \frac{\sin \Omega\tau}{\Omega\tau}.$$

We know that multiplication in the frequency domain is the same as convolution in the time domain, thus for the functions shown

$$R_{\Omega xx}(\tau) = \int_{-\infty}^{\infty} e^{-a|\tau|} \frac{\Omega}{\pi} \frac{\sin \Omega(\eta - \tau)}{\Omega(\eta - \tau)}\,d\eta.$$

We will not attempt to evaluate this but the result is 'lobe error' on the auto-correlation function.

[We could, of course, work from first principles and put

$$R_{\Omega xx}(\tau) = \frac{1}{2\pi} \int_{-\Omega}^{\Omega} \frac{2a}{a^2 + \omega^2} e^{j\omega\tau}\,d\omega$$

and simplify a little by using the fact that $\dfrac{2a}{a^2 + \omega^2}$ is an even function and obtain

$$R_{\Omega xx}(\tau) = \frac{2}{\pi} \int_{0}^{\Omega} \frac{a}{a^2 + \omega^2} \cos \omega\tau\,d\omega$$

but even this is not a simple integration for manual evaluation. It is, of course, straightforward on a computer for specified values of the parameters.]

6.11   Illustration

The results obtained for Figs. 6.16, 6.18, 6.19, 6.20 and 6.22 were produced using a Bartlett window at the Transform stage. However, Fig. 6.29 shows S_{yy} produced using a rectangular window and the 'side lobes' are very pronounced which would produce incorrect interpretation if the effect was not accounted for.

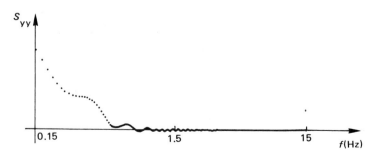

Fig. **6.29**

Exercises 6

1 A first order system has an impulse response given by

$$g(t) = 4e^{-2t} \qquad \text{for } t \geqslant 0$$

$$= 0 \qquad \text{for } t < 0$$

and is excited by a cosine signal $u(t) = 5 \cos 2t$ which was applied in the infinite past. Use frequency domain methods to find the magnitude and phase shift of the output. Compare with the results obtained in Question 2, Chapter 5.

2 A system whose impulse response is given by

$$g(t) = \tfrac{1}{2}(e^{-2t} - e^{-4t}) \qquad \text{for } t \geqslant 0$$

$$= 0 \qquad \text{for } t < 0$$

undergoes a frequency response test using a suitable range of frequencies. Using the graphical method of section 6.3 obtain frequency response data and plot as a pair of Bode diagrams.

3 Using any method plot the frequency response functions of the stable systems listed in Question 9, of Chapter 5.

4 Form $G(z)$ for (a) the system in question 1, and (b) that in Question 2, using a sampling interval $\Delta = 0.05$. For both systems use the method of section 6.6 to obtain frequency response data and plot as Bode diagrams. Show that the responses are periodic in ω.

5 Repeat the two parts of Question 4 when each system is preceeded by a zero-order hold.

6 For the signal and system described in Question 10, Chapter 5, find the cross-spectral density $S_{uy}(\omega)$. Is this result consistent with the result obtained in Question 10, Chapter 5? Also find $S_{yy}(\omega)$—is this result consistent? (Inverse Fourier Transform to check).

7 The system whose impulse response is given by

$$g(t) = \tfrac{1}{2}(e^{-2t} - e^{-4t}) \qquad \text{for } t \geqslant 0$$

$$= 0 \qquad \text{for } t < 0$$

is driven by a random signal whose auto-correlation function is $R_{uu}(\tau) = 200\, e^{-20|\tau|}$. Find $S_{uy}(\omega)$ and $S_{yy}(\omega)$. Are these results consistent with those obtained in Questions 13 and 14 of Chapter 5.

8 The result of a cross-correlation test using White Noise as input to a system yields a cross-correlation function estimated to be

$$R_{uy}(\tau) = 40\, e^{-2\tau}; \qquad \tau \geqslant 0;$$

$$= 0 \qquad ; \tau < 0$$

In order to estimate its frequency response a Fourier Transform is performed to obtain

$S_{uy}(\omega)$ (cf. section 6.7) but only a limited range for τ is possible because, theoretically, τ extends to infinity. Let the range for τ be a so that $R_{uy}(\tau)$ is effectively windowed (rectangular). Show that $S_{uy}(\omega)$ is

$$|S_{uy}(\omega)| = 40\sqrt{\left[\frac{(1-e^{-2a}\cos\omega a)^2 + e^{-4a}\sin^2\omega a}{4+\omega^2}\right]}$$

$$\angle S_{uy}(\omega) = \text{Arctan}\left(\frac{e^{-2a}\sin\omega a}{1-e^{-2a}\cos\omega a}\right) - \text{Arctan}\left(\frac{\omega}{2}\right)$$

Hence demonstrate that for $a \to \infty$, $S_{uy}(\omega)$ tends to the correct value. Use the Wiener–Kinchine relationship not convolution.

9   In an attempt at obtaining a better estimate of the frequency response of the system in Question 8, a Hanning window, $w(\tau) = \dfrac{1}{2}\left(1 + \cos\dfrac{\pi\tau}{a}\right)$ was used. Show that $S_{uy}(\omega)$ is

$$S_{uy}(\omega) = 20\left[\frac{1}{(2+j\omega)} + \frac{(2+j\omega)}{4+4j\omega+\dfrac{\pi^2}{a^2}-\omega^2} - \frac{\dfrac{\pi^2}{a^2}e^{-2a}}{(2+jw)\left(4+4j\omega+\dfrac{\pi^2}{a^2}-\omega^2\right)}(\cos\omega a - j\sin\omega a)\right]$$

Note that as $a \to \infty$ $S_{uy}(\omega) \to \dfrac{40}{2+j\omega}$.

10   Make computer programmes to plot out $|S_{uy}(\omega)|$ and $\angle S_{uy}(\omega)$ for the results of both Questions 8 and 9. That for 8 is straight forward but that for 9 is not so and the expression will need to be modified. One way would be to arrange the result in Real and Imaginary form, $(A+jB)$, and then compute $\sqrt{A^2+B^2}$ and $\arctan\dfrac{B}{A}$ for different ω.

7

The Measurement and Computation of Signal Characteristics

7.1 Introduction

In the preceding chapters we have considered the background to the characterisation of signal properties. In all cases both the analogue (continuous) and the digital (discrete) formula or algorithm have been presented. The term algorithm is used here rather than 'discrete formula' since the implementation of such an equation will be via a digital machine, which can either be a general purpose computer or a special purpose digital instrument.

Analogue methods of measuring signal characteristics are normally implemented using dedicated hardware but most of these instruments have been replaced by the digital equivalent. The design of modern instrumentation for signal processing is now microprocessor based which implement the algorithms mentioned above. For this reason most of this chapter is devoted to the digital aspects of measuring signal characteristics. In particular we will consider in detail the DFT and its application to the computation of frequency spectra.

The DFT has assumed pride of place among the formulae used in signal processing following the development of the Fast Fourier Transform (FFT). The algorithm for the fast computation of the DFT was published in 1965[3]. There then followed a period during which many techniques were developed for signal processing applications based on the FFT. This period was subsequently followed by the development of the microprocessor and together these two significant events have led to a new era in signal processing.

Following a review of analogue methods for measuring signal characteristics and a recap of the important formulae used in the evaluation of amplitude and time domain characteristics the majority of the chapter is devoted to digital spectral analysis. We will now look at the implications of using the DFT since the spectra which result from this algorithm can be completely erroneous unless some insight is gained into the significance of the DFT in relation to the analysis of real signals. The understanding of the DFT is of much greater importance than the FFT despite the obvious significance of the FFT. The FFT should be viewed as a tool, the results from which can be improved by reading the instructions, in this case the DFT.

These few words of caution should not deter the reader from using the DFT to obtain spectra but are intended to draw attention to the possible pitfalls into which the unwary may stumble when using these techniques for the first time. This also applies to the use of special purpose digital spectrum analysers. These instruments are designed to offer the maximum flexibility to the user, thus providing a number of options for analyses. Although the manuals offer advice on the options to be used for a particular application, the background information necessary to supplement this advice and hence enable the user to understand the significance of the options is not readily available. This chapter is intended to bridge this gap.

7.2   Analogue Methods of Measurement

Analogue instrumentation for the measurement of signal characteristics is now rapidly becoming obsolete. However, for the purposes of completeness and to reinforce some of the basic concepts introduced in earlier chapters we will briefly review the most common types of instrument which are based on analogue principles.

7.2.1   Measurement of Amplitude Characteristics

The primary parameters associated with the amplitude characteristics of signals are; the mean, r.m.s. and mean squared values together with the cumulative probability distribution (CPD) and the probability density distribution (PDD). Of these parameters the r.m.s. level of a signal is typically measured using analogue instrumentation. In this case an r.m.s. voltmeter would be used to determine the effective power in a record of a particular length. If the r.m.s. level is known then of course the mean squared value can also be determined.

In order to measure the distribution functions it is more convenient to use digital techniques since by definition these functions require digital information concerning the signal. This aspect will be discussed in section 7.3. However, it is possible to measure the PDD using analogue circuits.

7.2.2   Measurement of Time Domain Characteristics

Of the time domain characteristics only the auto-correlation function has been measured using analogue techniques. Indeed, as in the case of amplitude characteristics only a small range of special purpose instruments have been available until the introduction of digital instruments. One of the main reasons for this, in the case of correlation functions, is the design of suitable delay circuits whereas discrete signals can be delayed very easily in a digital system.

The diagram for the design of an analogue correlator is shown in Fig. 7.1. Such an instrument would have a fixed number of delays and so the usefulness of the instrument would be limited due to the lack of flexibility in the design. The diagram of a typical correlator does, however, serve to illustrate the realisation of the integral equation (3.10) in terms of hardware. The signal is delayed by a predetermined amount, multiplied by the original signal and the product integrated over some fixed time interval. This operation is repeated for as many time delays as the instrument allows. Such an instrument is very expensive since each delay must have its own multiplying and integrating circuit, as there are no means of storing the signal which enables each set of circuits to be used in turn.

7.2.3   Measurement of Frequency Domain Characteristics

Spectrum or frequency analysers have been available for a considerable length of time. One design of an analogue analyser is in fact based on a very simple principle which demonstrates the concept of frequency analysis. Again electrical circuits are required; in this case the principle element is a band-pass filter which has the frequency characteristics shown in Fig. 7.2. Thus only those frequency components of the signal which lie in the range f_1 to f_2 will be unaffected by the filter, all other components will be significantly reduced in amplitude.

If the output from such a filter is then connected to an r.m.s. voltmeter a measure of the amplitude of the components within the frequency band f_1 to f_2 is obtained. In order to

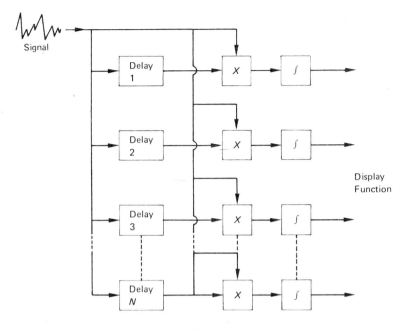

Fig 7.1

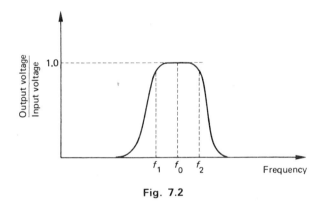

Fig. 7.2

obtain the frequency characteristics over a range of frequencies then either a number of filters and r.m.s. voltmeters must be arranged in parallel, as shown in Fig. 7.3, or the centre frequency (f_0) of the filter must be capable of being tuned to a range of values.

As one can imagine the cost of such an analyser, of either the fixed or variable frequency type, depends on the degree of sophistication involved in the design of the filter. Ideally the filter should have a sufficiently narrow bandwidth [($f_2 - f_1$ defines the bandwidth)] to be capable of discriminating between very small changes in frequency. However, consider the effect of a small bandwidth on the amplitude of the r.m.s. output voltage from the filter. As the bandwidth is reduced so the r.m.s. level will be reduced, hence the amplitude of the resulting spectrum will be dependent on the filter bandwidth. In practice if the bandwidth of the filter is halved then the power associated with the output from the filter is halved and

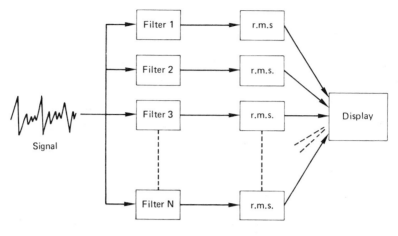

Fig. 7.3

since this power is proportional to the square of the r.m.s. voltage (the root mean squared value), on average the r.m.s. voltage will drop by $1/\sqrt{2}$. In order to overcome this problem we must divide by some factor associated with the filter bandwidth so that the resulting value is independent of the filter characteristics. If we divide by (filter bandwidth)$^{1/2}$ then the units of the resulting value will be r.m.s. volts/(hertz)$^{1/2}$ which is of course the square root of the units for power spectral density. Hence, once again, we see that the power content of a signal is the most suitable parameter to use when quantifying spectral information.

A bandpass filter of infinitely small bandwidth is unattainable and so resolution has to be sacrificed for design considerations. An alternative approach is to use the heterodyning principle.[18]

7.3 Digital Methods

The remainder of the chapter is primarily concerned with the computation of the discrete Fourier transform (DFT). The DFT forms the basis of digital spectrum analysers and can also be used to obtain other signal characteristics (correlation functions) and system characteristics (transfer function and, impulse response via indirect convolution). The theoretical basis for the DFT has been established in Chapter 4. Our intention here is to investigate the implications of using the DFT for signal characterisation.

Before investigating the DFT it is worth restating the discrete equations for the amplitude and time characteristics. The application of these equations has been discussed in Chapters 1, 2 and 3 and so only the equations will be given.

$$\text{Mean value } \tilde{x} = \frac{1}{N+1} \sum_{i=0}^{N} x_i \tag{7.1}$$

$$\text{Mean squared value } \widetilde{x^2} = \frac{1}{N+1} \sum_{i=0}^{N} x_i^2 \tag{7.2}$$

$$\text{Auto-correlation function } R_{xx}[k] = \frac{1}{N+1-k} \sum_{i=0}^{N-k} x_i x_{i+k} \tag{7.3}$$

$$\text{Cross-correlation function } R_{xy}[k] = \frac{1}{N+1-k} \sum_{i=0}^{N-k} x_i y_{i+k}. \tag{7.4}$$

An indirect method of computing correlation functions is to evaluate the appropriate spectral density function, via the DFT, and then to apply the inverse DFT to this function to determine the required correlation function (section 4.8). The correlation function which results is known as a circular correlation function. Such a function also results when the DFT is used to implement convolution. An explanation of circular convolution is given in Chapter 8 together with a method of overcoming the problems associated with circular convolution. A similar approach can be adopted to avoid circular correlation.

7.4   The Discrete Fourier Transform (DFT)

We have seen in Chapter 4 that the discrete equivalent of the Fourier transform is given by equation (4.13) and application of this equation to a time series results in the computation of discrete spectral coefficients. It was also shown in Chapter 4 that the Fourier transform is an alternative way of representing the complex coefficients of the Fourier series for a time limited signal. Thus, the DFT is a means of computing the discrete Fourier coefficients of a time series which is assumed to exist for a finite period; such a time series is shown in Fig. 7.4. The period of this finite (or truncated) record is *T*. Now, if this record were to be analysed via the Fourier series then the record would be assumed to be periodic and hence repetitive from $-\infty$ to $+\infty$. It is therefore reasonable to use the Fourier series/DFT analogy to draw the conclusion that by implementing the DFT on a truncated time series, the time series is itself assumed to be periodic: the period of repetition being the observation time *T*. At this stage such a deduction may appear to be only of academic interest. This is not so and the effects of this periodicity can be observed as soon as we move away from the analysis of ideal signals.

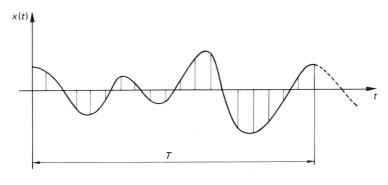

Fig. 7.4

If the application of the DFT infers that a time series is periodic, is it also possible that the resulting spectrum is periodic? This is of course, a new question as the Fourier series representation of a signal often results in an infinite series with a frequency spectrum extending to infinite frequency. However, we must now consider what happens if we try to compute spectral coefficients outside the normal range of $x_i(i=0, 1 \ldots N-1)$ say $i > N-1$ and $i < 0$.

7.4.1   The Periodicity of the DFT

The DFT defined by equation 4.13 is shown here with the scale factor $1/N$ omitted for convenience.

$$X_n = \sum_{k=0}^{N-1} x_k e^{-j2\pi kn/N}. \tag{7.5}$$

The transformation of the time series x_k therefore yields N spectral coefficients if equation (7.5) is applied N times, i.e. n takes the values 0 to $N-1$. Now consider the computation of a coefficient beyond the $(N-1)$th, say the $(N+n)$th coefficient. Then,

$$n \text{ becomes } N+n$$

and substituting for n into equation (7.5)

$$X_{N+n} = \sum_{k=0}^{N-1} x_k e^{-j2\pi k(N+n)/N}. \tag{7.6}$$

Rearranging the exponential term on the right hand side of equation (7.6)

$$X_{N+n} = \sum_{k=0}^{N-1} x_k e^{-j2\pi kn/N} . e^{-j2\pi k} \tag{7.7}$$

k is an integer

therefore $\quad\quad\quad\quad e^{-j2\pi k} = 1$

thus $\quad\quad\quad\quad X_{N+n} = \sum_{k=0}^{N-1} x_k e^{-j2\pi kn/N}$

$$X_{N+n} = X_n \tag{7.8}$$

so that each spectral coefficient is repeated every Nth coefficient and the DFT thus yields a periodic amplitude spectrum as shown in Fig. 7.5.

Now consider the case of coefficients in negative frequency (X_{-n}) and again substituting for n into equation (7.5).

$$X_{-n} = \sum_{k=0}^{N-1} x_k e^{-j2\pi k(-n)/N}$$

$$= \sum_{k=0}^{N-1} x_k \{\cos(2\pi kn/N) + j\sin(2\pi kn/N)\}. \tag{7.9}$$

The coefficient is thus the complex conjugate of X_n. This is written as

$$X_{-n} = X_n^*.$$

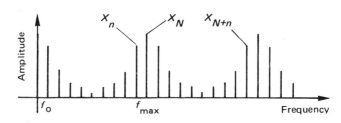

Fig. 7.5

The DFT therefore yields an amplitude spectrum which is also periodic in negative frequency, since

$$|X_n| = |X_n^*|.$$

Note that we have used the amplitude spectrum to validate the periodicity of the spectra thus losing phase information. This is perfectly acceptable as the spectrum of a signal is normally presented in terms of spectral density which as we have seen, depends only on the amplitude of the Fourier coefficients.

7.4.2 The Ordering of the Spectral Coefficients

The N spectral coefficients obtained by applying equation (7.5) N times to an N point time series yields an amplitude spectrum which is symmetric about the $N/2$ coefficient. To illustrate this consider the computation of the Fourier coefficient X_n where

$$n = N - r$$

and restrict r such that

$$r < N/2$$

The coefficient to be computed is in the second half of the spectrum. Now substitute for n into equation (7.5)

$$X_{N-r} = \sum_{k=0}^{N-1} x_k e^{-j2\pi k(N-r)/N}$$

$$= \sum_{k=0}^{N-1} x_k e^{j2\pi kr/N} . e^{-j2\pi k}$$

$$= \sum_{k=0}^{N-1} x_k e^{j2\pi kr/N}.$$

Thus $\qquad\qquad\qquad X_{N-r} = X_r^*.$ $\qquad\qquad\qquad\qquad\qquad$ (7.10)

Therefore the Fourier coefficients in the second half of the spectrum are the complex conjugate of those in the first half and the amplitude spectrum is symmetric about the mid-point as shown in Fig. 7.6.

In the previous section we noted that the spectral coefficients in negative frequency (X_{-n}) were also the complex conjugate values of the positive frequency coefficients. Therefore the

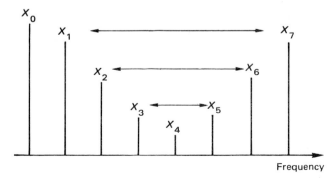

Fig. 7.6

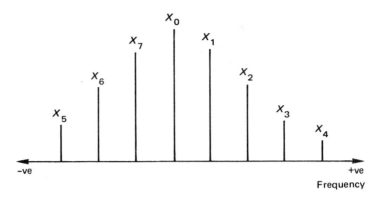

Fig. 7.7

second half of the computed spectrum is equivalent to the negative part of the spectrum. Thus the DFT yields a spectrum with a frequency range from $-\omega_{max}$ to $+\omega_{max}$ following the re-ordering of the computed spectrum, i.e. the amplitude spectrum shown in Fig. 7.6 is re-ordered as shown in Fig. 7.7.

Note that since X_0 is the steady or d.c. value (zero frequency value) there are $N/2$ points in the positive frequency range and $(N/2-1)$ points in the negative frequency range. The spectra associated with some typical signals are shown in Fig. 7.8.

The question must now be asked; what are the implications of this periodicity in the frequency domain as far as spectral analysis is concerned? It is often convenient when considering idealised signals such as the sine wave, to ignore the possibility of higher frequency components which may exist beyond the normal range of analysis. We will see in the next section that the frequency resolution of a DFT analysis is $1/T^*$ Hz (T as defined in Fig. 7.4) but this can be inferred by analogy between the Fourier series and DFT, i.e. the fundamental frequency component is $[(1/\text{Time period})]$ expressed in cyclic frequency (Hz). The total range of the frequency analysis is therefore $(N/(T))$ Hz for N spectral coefficients; but only the first half of the computed spectral coefficients are unique and hence the useful positive frequency range is from 0 to $(N/2T)$ Hz

$$\text{i.e. } f_{max}=(N/2T)\text{Hz} \quad \text{or} \quad \omega_{max}=(\pi N/T)\text{rad s}^{-1}.$$

Now consider the frequency analysis of a square wave with a fundamental period of 10 ms fundamental frequency of 100 Hz. The odd harmonics of this signal will be (100, 300, 500, 700 Hz etc.). If the analysis is done using the DFT for a one second time record (100 cycles) containing 500 samples then the frequency resolution will be 1 Hz and the frequency range from 0 to 250 Hz. This analysis will therefore only yield information about the fundamental component of the signal. The other higher frequency components have not disappeared, they are of course outside the range of the analysis. Now the symmetry of the DFT means that the spectrum within the valid range is repeated from 250 Hz to 500 Hz and this frequency range includes a further two spectral components, i.e. the 3rd and 5th harmonics. The effect of this is that because the periodic nature of the discrete spectrum dictates that successive frequency ranges are identical then the valid range from d.c. to 250 Hz must contain information about higher harmonics. This information appears as

---

* Note that the observation period of a signal can be either $-T$ to T (period $2T$) or $-T/2$ to $T/2$ or 0 to T (period T). The frequency resolution is adjusted accordingly.

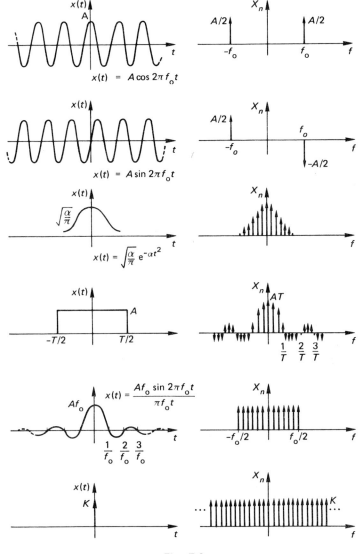

Fig. 7.8

erroneous spectral coefficients, see Fig. 7.9(b). These erroneous lines are said to have 'folded down' into the normal frequency range. The folding frequency, called the Nyquist frequency, is half the sample frequency, with corrupted coefficients retaining their correct spectral location with respect to the normal spectrum. This corruption of a spectrum is due to the phenomenon known as aliasing as discussed in Chapter 6 and is a result of under sampling the signal. Now use 500 samples but now observe the signal over a much shorter period of time and then see what happens to the frequency range.

Let $T = 0.1$ seconds

$$f_{max} = \frac{N}{2T} = \frac{500}{2 \times 0.1} = 2500 \text{ Hz}.$$

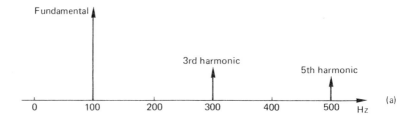

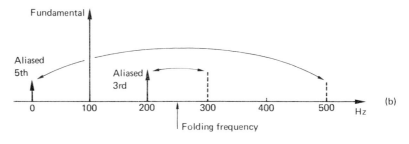

Fig. 7.9

In this case the valid frequency range is from d.c. to 2.5 kHz and so we would observe all the harmonics up to the 25th within the valid frequency range.

The parameter of interest is of course the sample rate. In the first example the sample rate was 500 Hz whereas for the second example the sample rate was 5 kHz. The spectrum of the square wave will still be aliased when the sample frequency is 5 kHz but because the amplitude of the 25th harmonic and above is small, (i.e. less than 4% of the fundamental component), then the effect of the aliasing will be negligible.

When the spectral characteristics of the signal are known in advance then the effects of aliasing will be immediately apparent, but when the signal is of unknown spectral content then care must be exercised. A solution to the problem is to ensure that an adequate sample frequency is used. Shannon's sampling theorem requires that in order to avoid aliasing the signal should be sampled at a rate which is at least twice the highest frequency present in the signal. To ensure that this requirement is satisfied to within acceptable limits the analogue time signal should be passed through a low pass filter prior to the sampling process. These filters are often called anti-aliasing filters, for obvious reasons, and are designed to have a sharp cut-off frequency with good attenuation outside the pass-band.

For a typical application the filter characteristics would be chosen so that the highest frequency within the pass-band (f_1) is approximately half the maximum frequency within the analysis range, as shown in Fig. 7.10.

Note that there will still be an aliased region but since the amplitudes of the spectral coefficients within this frequency range will be small the effect on the spectrum is not too significant. Many commercial analysers discard the top end of the computed frequency range and display, say, 400 out of 512 valid spectral components. This is done so that the effects of aliasing are minimised and the portion of the spectrum which is displayed is free from spurious components.

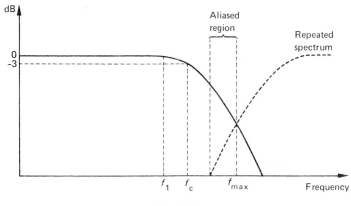

Fig. 7.10

7.4.3   Comment

We have seen in this section that the application of the DFT infers that both the time series and the resulting spectrum have the following properties.

(i)   The truncated time series is periodic.
(ii) The computed spectral coefficients are folded about the maximum frequency with the negative half of the spectrum appearing after the positive coefficients but in reverse order in Fig. 7.11.
(iii) The spectrum is periodic (in both positive and negative frequency). The spectrum shown in Fig. 7.11(b) is shown in Fig. 7.12 on an extended frequency scale to illustrate this.

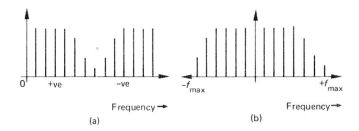

Fig. 7.11

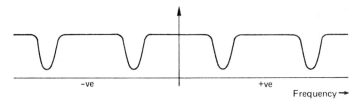

Fig. 7.12

The result of this periodicity is that an aliased time series will produce an aliased spectrum which results in erroneous spectral coefficients appearing in the valid frequency range.

7.4.4 The DFT applied to Real Signals

The examples given in the previous section show the discrete spectra of ideal signals. Consider now the truncated sinusoid shown in Fig. 7.13 which does not have an integral number of cycles within the observation period and starts with a non-zero magnitude.

We know that in reality the signal exists over an infinite time period (as indicated by the dotted lines) but for the purposes of analysis only the truncated portion of period T is available. The amplitude spectrum of such a signal is shown in Fig. 7.14.

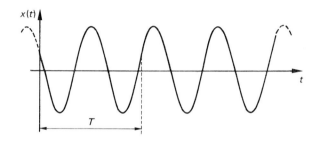

Fig. 7.13

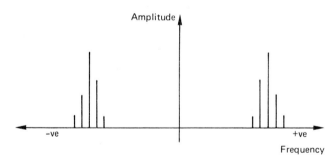

Fig. 7.14

We know that the signal is a sinusoid, but the spectrum indicates that other harmonics are present. This appears to imply that the DFT only gives the correct result for ideal signals; this statement is both true and false. The DFT gives the correct spectrum for the signal analysed but we have ignored two important points. Firstly, we have already seen that applying the DFT to a truncated time series infers that the signal is periodic. Thus the truncated signal is assumed to be part of the repetitive signal shown in Fig. 7.15.

This waveform is clearly different from the assumed form of sinusoid and if we had analysed this signal (now of period $3T$) then the spectrum shown in Fig. 7.14 would be plausible.

The second important point is that by truncating the signal an implicit mathematical operation has been performed on it prior to the DFT. An infinite sinusoid has been

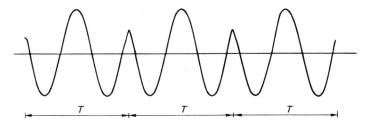

Fig. 7.15

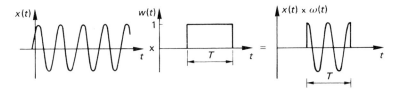

Fig. 7.16

multiplied by a function of unit amplitude and period T in order to yield the truncated sinusoid Fig. 7.16. This process is known as windowing and the function which operates on the signal is called a window function (see Chapter 6). In Fig. 7.16 a rectangular or uniform window function has been used. Indeed, this is the assumed form of window function whenever a signal is truncated.

The windowing operation has important implications when applying the DFT since the effect of multiplication in the time domain is to cause the spectra of the signal and the window function to be convolved in the frequency domain (see Chapter 6). This operation is shown in Fig. 7.17. The spectrum of the rectangular window is the well known envelope of the ($\sin x/x$) type of function. This is shown in more detail in Fig. 7.18. We can see from the

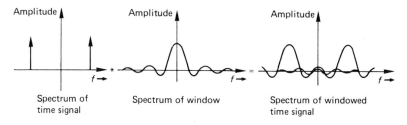

Fig. 7.17

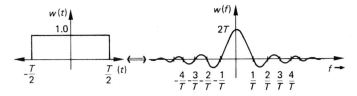

Fig. 7.18

spectrum of the window that at frequencies $(\pm 1/T, \pm 2/T, \pm \ldots \pm n/T)$ the amplitude of the spectrum is zero

$$\text{i.e.} \quad W(f)=0 \quad \text{when } f = \pm \frac{k}{T} \quad k = 1,2,3, \ldots, n$$

$$\text{or} \quad W(\omega)=0 \quad \omega = \pm \frac{2\pi k}{T}.$$

The width of the central lobe is thus $2/T$ Hz and is therefore a function of the period of the window. Thus, the longer the period of the window (the longer the time record) the narrower the central lobe becomes and in the limit (as T goes to infinity) the central lobe becomes infinitely narrow and so the effects of windowing become negligible.

The DFT of the truncated sine wave now looks even more plausible and the window function appears to explain the differences between the spectra of an ideal sine wave and that of a truncated sine wave. We will return to this problem after exploring another aspect of windowing. So far we have seen how the spectrum of a signal is modified by the data window but there is a further important consideration when analysing complicated waveforms with closely spaced harmonic components. To demonstrate the effect of windowing on frequency resolution consider the idealised spectrum shown in Fig. 7.19(a), (2 spectral components ω_1 and ω_2 with only positive frequencies shown for convenience).

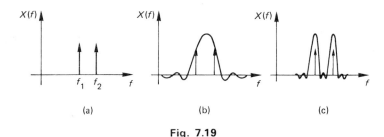

Fig. 7.19

If a window function with a short time period is used, (i.e. a severely truncated signal), then the width of the central lobe associated with the spectrum of the window will be sufficiently wide to mask the spectral components of the signal. The resulting spectrum will be as shown in Fig. 7.19(b) where the two components are merged into the central lobe. Increasing the period of the window reduces the width of the central lobe so that now both of the frequency components can be resolved as shown in Fig. 7.19(c). Thus the period of the window determines the frequency resolution of the analysis. The frequency resolution is determined by the width of the lobe of the window function, i.e.

$$\text{frequency resolution} = \Delta f = \frac{1}{\text{period of window } (T)} \text{ Hz.}$$

Although the adjacent spectral components can be resolved by the correct choice of window parameters they are, however, contaminated by the side lobes of the window function. Thus the observed spectral components are in error. The adjacent components are said to have 'leaked' into their respective neighbours and so this process is termed 'leakage'.

The effect of truncating a signal obviously has an important bearing on the results from a DFT analysis of the signal. To gain a better appreciation of the significance of windows

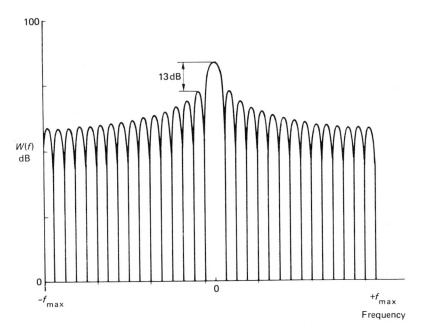

Fig. 7.20

consider the amplitude characteristics of the rectangular window plotted on a dB amplitude scale, Fig. 7.20. We see that the main lobe is 13 dB above the side lobes and the roll-off in the amplitude of the side lobes is 6 dB per octave (on a log frequency scale). Each spectral component is similarly weighted and so the cumulative effect can be significant and the apparently corrupted spectrum shown in Fig. 7.14 now looks like a plausible result.

How can the effects of windowing be reduced? This is an obvious question to ask at this stage and of course there are ways of improving the reliability of the results. Clearly attention must be focused on the window function in order to make any improvements. The requirements for an ideal window are that the central lobe should be narrow, to give good frequency resolution and the side lobes should be of insignificant magnitude in relation to that of the central lobe in order to reduce the effects of leakage. These requirements often conflict and so window functions have been developed which have the desired characteristics for a particular application[4, 5].

One function which offers a good compromise between frequency resolution and reduced leakage is known as the Hanning window. This window function is incorporated into many of the commercial digital spectrum analysers on the market. The Hanning window [see Fig. 7.21a] is defined by the mathematical expression

$$w(t) = \frac{1}{2}\left(1 + \cos\frac{2\pi(t - T/2)}{T}\right)$$

The spectrum for this function is shown in Fig. 7.21(b) with the amplitude again in dB. In this case the roll-off of the side lobes is 18 dB per octave on a log scale.

Another window function which is used in the analysis of random signals is the cosine taper window shown in Fig. 7.22(a). This function has the frequency characteristics shown in Fig. 7.22(b). From an inspection of Figs 7.21(b) and 7.22(b) we see that the cosine taper

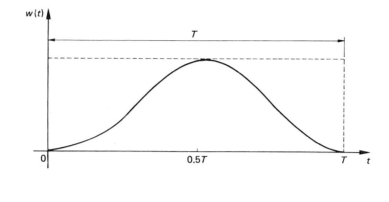

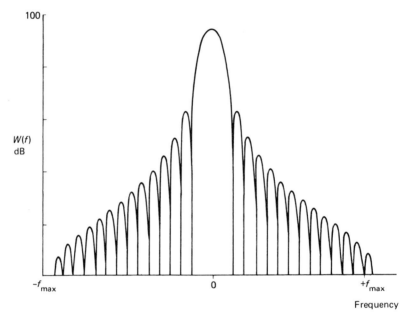

Fig. 7.21

window has a narrower central lobe than the Hanning window and therefore exhibits better frequency resolution. The Hanning window, however, has a better roll off rate and therefore reduces the leakage effect. This means that for a periodic or almost periodic signal the Hanning window will yield the best results since the repetitive nature of such signals causes maximum distortion of the spectrum due to windowing. This is demonstrated in Figs. 7.23 and 7.24 where a sine wave has been analysed using first a cosine taper and then a Hanning window.

There are numerous window functions, each of which have some desirable feature for some particular application. One important point to remember is that window functions, except the rectangular window, will weight the amplitude of each spectral coefficient and that if a particular application requires an accurate measure of amplitude then some care must be exercised when using window functions. Durrani and Nightingale[5] give figures for

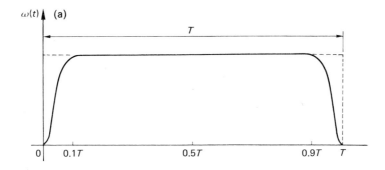

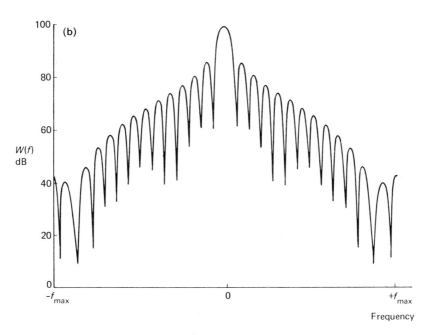

Fig. 7.22

the compensation which must be applied to the computed power spectrum, e.g.

$$G'(f) = G(f)/0.875 \qquad \text{for cosine taper window}$$

$$G'(f) = G(f)/0.375 \qquad \text{for Hanning window,}$$

where $G'(f)$ is the compensated spectrum in each case. For a more complete description of the relative merits of different data windows see Harris[4].

Before leaving data windows let us return to the sinusoid which, as we have seen, only yields the true spectrum if an integral number of cycles exist within the rectangular window. The reason for this now follows from the frequency characteristics of the rectangular window.

For an integral number of cycles the frequency of one of the computed spectral lines will be exactly equal to the frequency of the signal. The window function will therefore appear to be as shown in Fig. 7.25 where the magnitude of the window function is zero at every

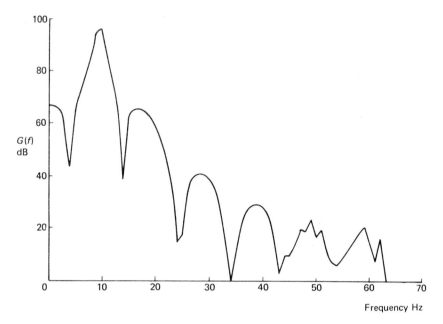

Fig. 7.23

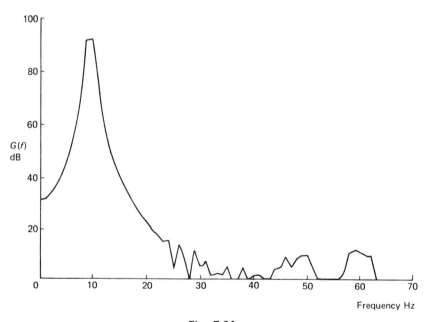

Fig. 7.24

spectral line but the central lobe. Hence, the analysis yields the expected single spectral line at the frequency of the sinusoid as shown in Fig. 7.26.

For a non integral number of cycles the window function will be as shown in Fig. 7.27(a) where the central lobe and the points of zero magnitude are not coincident with the computed spectral lines and the discrete spectrum appears as shown in Fig. 7.27(b).

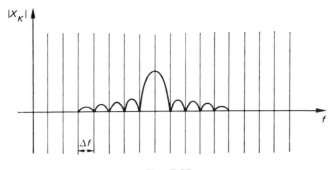

Fig. 7.25

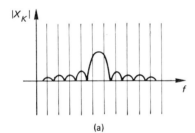

Fig. 7.26

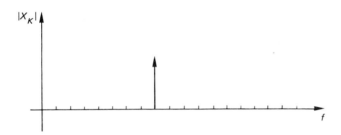

(a)   (b)

Fig. 7.27

If the analysis is performed on a waveform from a commercial signal generator then the sinusoid 'captured' within the data window will change from one test to another with the result that an apparently 'different' signal will be analysed each time. This in turn leads to a slight variation between the resulting spectra. If the spectra are now averaged these variations which tend to be random will be averaged out and a spectrum which approaches the desired single harmonic component will be achieved.

Another important aspect when analysing real signals, is that they may exhibit trends and characteristics which invalidate the ideas of stationary ergodic signals. Remember that if a signal is ergodic then the analysis can be performed on a time or ensemble basis and the results are equal. If the signal is clearly non stationary then no amount of adjustment will compensate for this but a typical instance where some initial pre-processing of data can improve the final result is when the signal exhibits a drift or slowly changing mean value.

The processing required to do this is termed trend removal. Note that if the trend appears to modify the signal in a cyclic manner then the trend should only be removed if the period is longer than the record length, otherwise the spectrum will be incorrect. Often the trend will be as a result of drift due to the limitations of instrumentation, in which case, an estimate of the compensating function will be straightforward but more complicated trends may pose a few difficulties.

Another useful pre-processing procedure is the removal of the steady or d.c. content, especially where the d.c. level is larger than the most significant peaks in the time-varying part of the signal, e.g. Fig. 7.28. In some cases the d.c. level should be adjusted to zero or some small level prior to sampling in order to reduce the effects of quantisation error on the low level variations. The removal of any offset from the sampled time series is straightforward since the mean value can be easily computed and then substracted from each data value in turn.

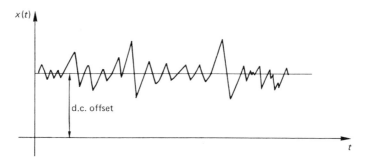

Fig. 7.28

For example, let the original time series be y_n, $n = 0, 1, 2 \ldots N$.

mean value
$$\tilde{y} = \frac{1}{N} \sum_{n=0}^{N-1} y_n.$$

The modified time series with zero mean value is then given by
$$x_n = y_n - \tilde{y}.$$

7.4.5   The Analysis of Random Signals

In all the preceding sections we have considered only deterministic signals and the implication of using the DFT to obtain their frequency spectra. In the majority of applications however, the signals under investigation will emanate from systems which give rise to random variations in the parameter under investigation. We must therefore, consider the effect of applying the DFT to the analysis of such signals.

The analysis procedure will be as previously outlined and all the constraints and precautions must be observed; the effects of aliasing must be minimised by the use of suitable low pass filtering prior to sampling and a suitable data window applied to the time series to reduce the effects of leakage. But now consider what happens if we perform the analysis on three arbitrarily selected sections of the stationary random waveform shown in Fig. 7.29(a) and compute the amplitude spectrum for each section using the same analysis

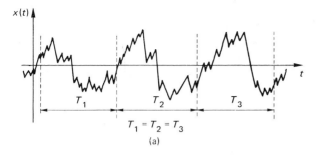

(a)

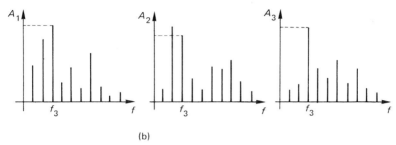

(b)

Fig. 7.29

procedure in each case, i.e. identical sample rates and number of samples. The spectra which result are shown in Fig. 7.29(b).

Now consider the amplitude of the frequency component f_3 in each of the three spectra. Quite clearly each amplitude is significantly different from the others and indeed each of the spectra exhibit different characteristics. But we know that the signal is stationary and assume it to be ergodic so that performing the analysis on any section of the waveform should produce representative results. It could be argued that if the length of each record were extended then the results would be more consistent but in practice it will only be possible to analyse a relatively short length of signal. Therefore an alternative means of improving the constancy of the spectrum must be developed.

Before considering the means of improving the reliability of the spectrum let us think about the result which may be expected from application of the DFT to random signals. The DFT can be written as

$$X_n = \frac{1}{N} \sum_{k=0}^{N-1} \{x_k \cos(2\pi kn/N) - jx_k \sin(2\pi kn/N)\} \tag{7.11}$$

or $X_n = X_{nR} - jX_{nI}$.

Thus when X_n is a random variable it is reasonable to expect that both X_{nR} and X_{nI} (the real and imaginary parts of a particular frequency component) will be random variables, since the Fourier transform is a linear operation operating on a random variable. This means, that computing the spectrum from a single time record of a random signal in fact produces a random result. Therefore, whatever processing is subsequently performed on the computed spectrum the best that can be hoped for is a 'good estimate' of the true spectrum. So from this point we will call each computed spectral component the estimated component.

The most obvious method of improving the reliability of the spectral estimates is to perform some type of averaging process. This can be done for each spectral coefficient in turn so that the coefficient of the spectra shown in Fig. 7.29(b) can be averaged at each frequency in turn. Thus, at frequency f_3 the averaged component is

$$\tilde{A} = \tfrac{1}{3}(A_1 + A_2 + A_3).$$

The averaged spectrum is then as shown in Fig. 7.30 and is an ensemble average.

Two types of spectra resulting from the analysis of random data deserve special mention. One is the 'flat' spectrum which results when all the frequency components within the analysis bandwidth have equal amplitude. This is known as 'white noise' by analogy with the definition of white light. The other is the class of spectra where the amplitudes of the spectral components 'roll-off' at a constant rate, e.g. 10 db/decade. This is often called coloured noise.

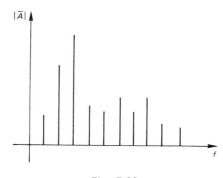

Fig. 7.30

Whilst the averaging of data to produce more reliable estimates is an accepted procedure with a sound statistical background (see Chapter 2) we must consider the validity of averaging spectral estimates and quantify the improvement in the estimates. Unfortunately this leads to some rather involved mathematics which is outside the scope of this book*. However, the important result of the analysis is that the ratio of the standard deviation to the true mean of the estimates is a constant for a particular set of analysis parameters and is given by

$$\frac{\sigma}{\mu} = \frac{1}{\sqrt{BT}} \qquad (7.12)$$

where σ = standard deviation;
         μ = mean value
         B = bandwidth of the analysis; (Hz)
         T = record length (seconds).

The bandwidth referred to here is in fact the frequency resolution of the DFT analysis (in analogue terms the bandwidth of the narrow band filter). But we know that the frequency resolution is related to the period of the record T,

i.e. $$B = \frac{1}{T}$$

* See Newland (6) or Bendat & Piersol (1).

thus
$$\frac{\sigma}{\mu} = 1$$

This means that the standard deviation of each spectral density estimate is equal to the mean (or expected) value. The estimate is therefore 100% uncertain.

In order to improve the reliability of the estimates we must somehow reduce this ratio. One approach to this problem which can be easily implemented is to increase the effective record length by analysing several time records of equal length (say L records) and take the ensemble average of these. The total record length then becomes $L \times T$ and hence

$$\frac{\sigma}{\mu} = \sqrt{\frac{1}{BLT}} = \frac{1}{\sqrt{L}}. \qquad (7.13)$$

We now require some means of assessing the confidence which can be placed on the estimates in order to determine the effectiveness of averaging. This can be done statistically from a knowledge of the probability density function (PDD) associated with the distribution of the estimates. For a normal distribution confidence limits can be determined from the standard deviation, e.g. if the mean value is 10 and standard deviation 1.5 then there is 66% certainty that any value would lie in the range 10 ± 1.5 (the mean value $\pm$ one standard deviation).

The computed spectral estimates form a χ_k^2 distribution (pronounced Keye square) where k defines the number of statistical degrees of freedom in χ_k^2. This distribution has a different PDD for each degree of freedom. Now the ratio of the standard deviation to the mean of a χ_k^2 process is given by

$$\frac{\sigma}{\mu} = \sqrt{\frac{2}{k}} = \frac{1}{\sqrt{L}}.$$

Thus
$$k = 2L \qquad (7.14)$$

From a knowledge of k the confidence limits can be established from tables of statistical parameters[7].

Example

The PSD of a signal is computed from the ensemble average of 20 time records. Determine the 90% and 98% confidence intervals. The number of degrees of freedom $k = 2 \times L = 40$. To establish the 90% confidence interval we must know the probability of the estimate lying between 5% and 95% of the expected mean value. These values can be looked up in the tables[6] and yield

$$1.508 \, \hat{G}_{xx} > \bar{G}_{xx} > 0.717 \, \hat{G}_{xx}$$

i.e. the expected value $\bar{G}_{xx}$ lies between 1.508 and 0.717 times the computed spectral estimate $\hat{G}_{xx}$. The 98% confidence limits are given by

$$1.691 \, \hat{G}_{xx} > \bar{G}_{xx} > 0.628 \, G_{xx}.$$

From these values it is clear that in order to obtain reliable spectral estimates it is necessary to average as many records as possible as there is still considerable uncertainty in the estimates even after averaging 20 records.

7.4.6   Parameters associated with the DFT

We have previously noted that once the sampling frequency and the number of samples have been selected then the range and resolution of the resulting spectrum are also selected. To underline this the important relationships are now given together for ease of reference.

$$\Delta = \frac{1}{f_s} = \left(\frac{2\pi}{\omega_s} \right) \qquad (7.15)$$

$$T = N.\Delta \qquad (7.16)$$

thus

$$\Delta f = \frac{1}{T}$$

or combining equations (7.15) and (7.16)

$$\Delta f = \frac{f_s}{N} = \left(\frac{\omega_s}{2\pi N} \right) \qquad (7.17)$$

where N is the number of samples
Δ is the sample interval
T is the total sample length
f_s is the sample frequency
Δf is the frequency resolution.

The frequency range is now given by

$$f_R = \Delta f . N = f_s = \left(\frac{\omega_s}{2\pi} \right) \qquad (7.18)$$

and the maximum and minimum frequencies are

$$f_{max} = \frac{f_s}{2} = \left(\frac{\omega_s}{4\pi} \right) \qquad (7.19)$$

$$f_{min} = \frac{-f_s}{2} = \left(\frac{\omega_s}{4\pi} \right). \qquad (7.20)$$

7.5   The Fast Fourier Transform (FFT)

The FFT is an extremely efficient procedure (algorithm) for the computation of the discrete Fourier transform. The necessity to have an algorithm which is computationally efficient can be appreciated by considering the DFT in more detail.

The DFT, as defined in Chapter 4, is given by

$$X_n = \frac{1}{N} \sum_{k=0}^{N-1} x_k e^{-j2\pi nk/N}. \qquad (7.21)$$

Applying this equation for a particular time series yields the complex coefficients X_n. Therefore to obtain N spectral coefficients equation (7.21) must be applied N times. Now consider the mathematical procedures involved in evaluating equation (7.21) for the most general case when x_k is assumed to be complex. Remember, in the transformation from frequency to time the coefficients will in general be complex but in the time to frequency transformation the time series, x_k, will normally be real. Writing x_k as a complex time series and expanding the exponential term into cosine and sine terms yields

$$X_n = \frac{1}{N} \sum_{k=0}^{N-1} [Rx_k + jIx_k]\left[\cos\left(\frac{2\pi nk}{N}\right) - j\sin\left(\frac{2\pi nk}{N}\right)\right] \tag{7.22}$$

or

$$X_n = \frac{1}{N} \sum_{k=0}^{N-1}\left[\left\{Rx_k\cos\left(\frac{2\pi nk}{N}\right) + Ix_k\sin\left(\frac{2\pi nk}{N}\right)\right\} + j\left\{Ix_k\cos\left(\frac{2\pi nk}{N}\right)\right.\right.$$
$$\left.\left. - Rx_k\sin\left(\frac{2\pi nk}{N}\right)\right\}\right]. \tag{7.23}$$

Thus to evaluate one complex spectral coefficient requires $4N$ multiplications and $2N$ additions. For N spectral coefficients this will mean $4(N)^2$ multiplications and $2(N)^2$ additions and if the coefficients are averaged over L time records then the amount of computation becomes prohibitive. For this reason the DFT is only used directly for small record lengths ($N \leqslant 100$). Before the publication of the FFT algorithm, frequency spectra were computed via the correlation function and the Weiner–Kinchine relationship, Blackman and Tukey[8]. After the publication of the FFT algorithm in 1965 the Blackman–Tukey method was no longer necessary and the DFT is now implemented via the fast algorithm.

We have already stated that the FFT should be regarded as a 'tool', albeit an extremely useful tool, for application to signal processing problems. The use of the algorithm does not require a detailed understanding of the procedures involved. This is true even for those wishing to incorporate the FFT into a computer program provided the algorithm is available as a subroutine or procedure. However, we feel that a book of this nature should include a review of this important algorithm but at first reading this can be overlooked without loss of continuity.

Consider the DFT of an N point sequence, now written as shown in equation (7.24), where the factor $1/N$ has been dropped for convenience.

$$X_n = \sum_{k=0}^{N-1} x_k W^{nk} \tag{7.24}$$

where

$$W = e^{-j2\pi/N}. \tag{7.25}$$

Now if N is divisible by 2 then the summation can be separated into summations over the even and odd indices of N respectively. Equation (7.24) now becomes

$$X_n = \sum_{k=0}^{N/2-1} x_{2k} W^{2nk} + \sum_{k=0}^{N/2-1} x_{2k+1} W^{(2k+1)n} \tag{7.26}$$

or

$$X_n = \sum_{k=0}^{N/2-1} x_{2k} W^{2nk} + W^n \sum_{k=0}^{N/2-1} x_{(2k+1)} W^{2nk}. \tag{7.27}$$

The first summation is the DFT of a sequence $e_0, e_1 \ldots e_{N/2-1}$ formed from the even indices of X_n, i.e.

$$E_n = \sum_{k=0}^{N/2-1} x_{2k} W^{2nk}. \tag{7.28}$$

Similarly, the second summation is the DFT of a sequence $o_1, o_2 \ldots o_{N/2-1}$ the odd indices of X_n, and

$$O_n = \sum_{k=0}^{N/2-1} x_{(2k+1)} W^{2nk}. \tag{7.29}$$

It is important to note at this stage that although the W factor has been raised by a power of 2, compared with equation (7.24) E_n and O_n are still valid DFT's of their respective sequences since

$$W^{nk} = e^{-j2\pi nk/N}.$$

Replacing N by $N/2$ gives

$$e^{-j2.2\pi nk/N} = W^{2nk}.$$

The DFT of the complete N point sequence is now given by substituting equations (7.28) and (7.29) into equation (7.27)

$$X_n = E_n + W^n O_n. \tag{7.30}$$

The DFT of E_n and O_n can each be computed with $(N/2)^2$ operations and X_n computed from a further N operation. Therefore the total number of operations to compute N spectral coefficients is

$$2\left(\frac{N}{2}\right)^2 + N.$$

If N is large then the saving in computational time becomes significant. However, the formulation of the procedure appears to limit any analysis to $N/2$ points since E_n and O_n are only computed over the range 0 to $(N/2-1)$. At this point the symmetry of the DFT, which can cause difficulties from the point of view of spectral analysis, comes to our aid.

Thus the spectral coefficients of each half sequence repeat themselves with period $N/2$ so that

$$E_{(n+N/2)} = E_n$$

$$O_{(n+N/2)} = O_n. \tag{7.31}$$

Therefore the second half of the complete spectrum $(n = N/2+1$ to $N-1$ can be computed from equation (7.30)

i.e

$$X_{n+N/2} = E_n + W^{n+N/2} O_n. \tag{7.32}$$

Now

$$W^{n+N/2} = e^{-j2\pi(n+N/2)/N}$$

$$= e^{-j2\pi n/N} . e^{-j\pi}$$

and noting that

$$e^{-j\pi} = -1$$

$$W^{n+N/2} = -W^n.$$

Therefore, equation (7.32) can be written as

$$X_{n+N/2} = E_n - W^n O_n. \tag{7.33}$$

The complete algorithm for the reduction of an N point DFT to two $N/2$ point DFT's can be shown diagrammatically, see Fig. 7.31. This is known as the 'butterfly' diagram.

Example

The reduction of an 8 point DFT to two 4 point DFT's is shown in Fig. 7.32. The time series of even and odd indices are shown opposite each 4 point DFT. If E_k and O_k can also be divided into two time series of equal lengths then a similar procedure can be adopted as in the case of the 8 point sequence, 4×2 point sequences can be transformed and recombined.

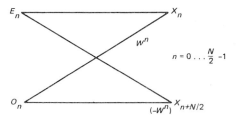

Fig. 7.31

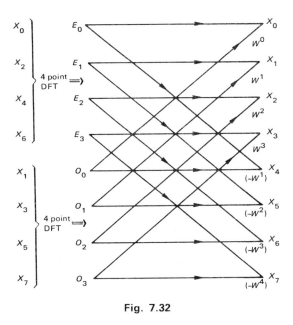

Fig. 7.32

Finally, each of the two point sequences can be reduced to a 2×1 point sequence. The DFT of a single point sequence is given by

$$X_n = \sum_{k=0}^{N-1} x_k e^{-j2\pi kn/N} \qquad n=0, \ k=0, \ N=1$$

$$X_0 = x_0.$$

The complete computational diagram for the reduction of an 8 point sequences to 8×1 point sequence is shown in Fig. 7.33. Provided N can be divided into N single point sequences the above procedure can be used. To ensure this $N = 2^m$. You may observe from Fig. 7.32 that in order to obtain the transform of the time series in the correct order ($n=0$, $1 \ldots N-1$) the time series must be re-ordered. Clearly because of the decomposition into odd and even series all the odd indexed terms will be in one half of the reordered sequence and the even indexed terms in the other half. As the reduction progresses so the original series becomes more disordered.

At this point the finesse of the complete algorithm can be appreciated. The reordering process can be seen to be a logical procedure if the index or address of each discrete time data point is expressed as a binary number. Then by reversing the order of each binary

Time Frequency

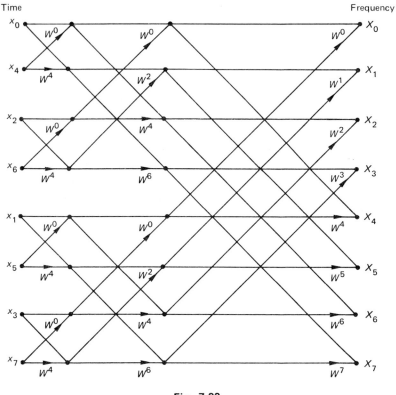

Fig. 7.33

index the correct order in which to start the FFT procedure can be determined. This initial re-ordering process is known as 'bit-reversal'. The reordering of the indices of the 8 point sequence is shown in Table 7.1 The first stage in the FFT algorithm is to re-order the time series sequence into bit reversed order. The most convenient way to do this is to use an assembly language program but high level procedures have been developed which are computationally efficient since the re-ordering process involves the swapping of two

Table 7.1

	Original Time Series		Recorded Time Series	
decimal index		*binary index*	*bit reversed index*	*decimal index*
x_0		x_{000}	x_{000}	x_0
x_1		x_{001}	x_{100}	x_4
x_2		x_{010}	x_{010}	x_2
x_3		x_{011}	x_{110}	x_6
x_4		x_{100}	x_{001}	x_1
x_5		x_{101}	x_{101}	x_5
x_6		x_{110}	x_{011}	x_3
x_7		x_{111}	x_{111}	x_7

numbers at a time in the series, e.g.

$$X_1 \text{ becomes } X_5$$

and
$$X_5 \text{ becomes } X_1.$$

FORTRAN and BASIC subroutines to implement the FFT algorithm are given in Appendix D. These subroutines perform the transformation procedure in both directions, i.e. time to frequency and frequency to time. The input sequence is assumed to be complex, thus the array holding the imaginary values must be set to zero prior to entry, for a real time sequence.

The speed of the FFT algorithm obviously depends on the computer used and the precise form of the algorithm implemented. Some further time saving 'dodges' are listed in the next section. The reduction of the time series to single points and the subsequent 'rebuilding' process requires $N \log_2 N$ complex operations. In addition to this the reordering process requires an additional N operations. To give some indication of the possible time savings consider the ratio of (DFT/FFT) operations given in Table 7.2.

Table 7.2

N	$\dfrac{\text{DFT}}{\text{FFT}} = \dfrac{N^2}{(N \log_2 N) + N}$
16	3.2
32	5.3
64	9.1
128	16.0
256	28.4
1024	93.1
4096	315.0

7.6   A Recipe for the Computation of Frequency Spectra

We now have all the ingredients necessary to compute reliable frequency spectra of real signals. The procedure is listed in the order in which a computer program would be written and follows the steps outlined in the flow chart shown in Fig. 7.34.

1   PSD

(a)   Sample the analogue time signal to produce the time series y_n—at this stage the sample frequency f_s and probably the number of data points N are selected. Thus the frequency range and resolution of the analysis will be defined.

(b)   Remove any d.c. offset and trends in the data, e.g. the effects of any drift which may be present. The mean value of the signal, $\tilde{y}$ must be computed so that a modified time series x_n with zero offset can be generated.

$$\tilde{y} = \frac{1}{N} \sum_{n=0}^{N-1} y_k$$

$$x_k = y_k - \tilde{y}$$

(c)   Apply a suitable data window to the time series to yield a spectrum with the required

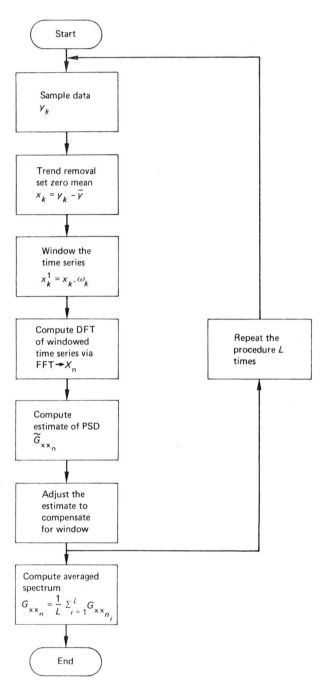

Fig. 7.34

frequency and/or amplitude resolution

$$x'_k = x_k \cdot w_k$$

where w_k is the weighting according to a prescribed window function.

(d)   Compute the DFT of the N point time series using the FFT algorithm. Note that the version of the FFT algorithm given in section 7.5 scales the time series x'_k by a factor $(1/N)$ so that some care should be exercised in drawing comparisons between formulae for the computation of PSD given here and those given in other texts.

$$X_n = \frac{1}{N} \sum_{k=0}^{N-1} x'_k e^{-j2\pi kn/N}$$

(e)   Retain only the first $N/2$ coefficients (the unique part) of the spectrum in the positive frequency range (0 to $f_s/2$) of X_n and compute the spectral density at each frequency component.

Remember $S_{Txx}(\omega) = \frac{1}{T}|X(j\omega)|^2$ (in continuous time).

Therefore by analogy in discrete time

$$\tilde{S}_{xx_n} = \frac{1}{T}|X_n|^2 = \frac{1}{T}X_n X_n^*. \tag{7.34}$$

However, in our derivation of the DFT we have already introduced a factor $1/T$ into the formula which is not present in the continuous form of the Fourier transform and therefore direct implementation of equation (7.34) will yield incorrect estimates of spectral density. Thus to yield the correct estimates equation (7.34) becomes

$$\tilde{S}_{xx_n} = N\Delta X_n X_n^*$$

where $N\Delta = T$.

This is the two sided spectral density but X_n has only been computed for the positive half of the frequency range since we are only interested in the single sided (positive frequency) spectral density. The single sided PSD is then given by

$$\tilde{G}_{xx_n} = 2N\Delta X_n X_n^*.$$

If W_f is the correction factor to compensate for the attenuation due to windowing, then the estimate is further modified

$$\tilde{G}_{xx_n} = 2W_f N\Delta X_n X_n^*.$$

(f)   The estimates computed are based on the transformation of a single time record and therefore for all but ideal signals are extremely poor estimates as shown in section 7.45. In order to improve the estimates we must average L spectra so that steps (a) to (e) must be repeated L times and the overall average computed. The number of averages will normally be chosen to give the required confidence levels for each estimate.

Thus $\hat{G}_{xx_n} = \frac{1}{L} \sum_{i=1}^{L} \tilde{G}_{xx_{ni}}.$

Note the units of the computed spectra are $[(\text{Amplitude})^2/\text{Hz}]$.

2   Cross Spectral Density

In this case there are two signals $x_1(t)$ and $x_2(t)$ to which steps (a) to (d) are applied to each signal in parallel. The sampling frequency must be the same for both signals. The single sided cross spectral density estimate is given by

$$\tilde{G}_{xx_n} = 2N\Delta X^*_{1n} X_{2n}.$$

This should be averaged in the same manner as the PSD to improve the reliability of the estimate.

3   Transfer Function

It was shown in Chapter 6 that the transfer function of a system is given by

$$G(f) = \frac{G_{xy}(f)}{G_{xx}(f)}$$

where $G_{xy}(f)$ is the single-sided cross spectral density function and
$G_{xx}(f)$ is the single-sided power spectral density function
$G(f)$ is the transfer function in terms of cyclic frequency (Hz).

Thus the best estimate of the transfer function for a system is given by the ratio of the cross and power spectral density estimates.

$$\hat{G}_n = \frac{\hat{G}_{xy_n}}{\hat{G}_{xx_n}}.$$

4   Ordinary Coherence Function

The ordinary coherence function is defined by

$$\eta^2(f) = \frac{G_{xy}(f)G^*_{xy}(f)}{G_{xx}(f)G_{yy}(f)}.$$

Since we are taking the ratios of terms scaled by the same factors the discrete equivalent can be substituted directly, thus

$$\eta^2_n = \frac{\hat{G}_{xy_n}\hat{G}^*_{xy_n}}{\hat{G}_{xx_n}\hat{G}_{yy_n}}.$$

The estimates must be averaged before computing the coherence function since the computation of n^2_n from a single estimate will always yield a coherence of unit viz

$$\eta^2_n = \frac{X^*_n Y_n X_n Y^*_n}{X^*_n X_n Y^*_n Y_n} = 1.$$

7.7   The DFT of Two Real Signals

In carrying out the computation of cross-spectra, transfer functions and coherence functions we require the transformation of two real signals from time to frequency. The two signals are normally the system input (or excitation signal) and the system output (the response). However, the DFT can be used to transform both of these signals in 'one-go', hence reducing computation time still further, the argument is as follows.

The DFT has been derived for complex signals but we noted that the imaginary part is zero in the case of real signals. Now consider the case when a signal is given by

$$a_k = x_k + jy_k$$

where x_k and y_k are individual real signals.

The DFT of a_k is then given by

$$A_n = \sum_{k=0}^{N-1} (x_k + jy_k)e^{-j2\pi kn/N}. \tag{7.35}$$

However, we require X_k and Y_k which can be determined by noting that

$$e^{j2\pi k(N-n)/N} = e^{j2\pi k} \cdot e^{-j2\pi kn/N} = e^{-j2\pi kn/N} \tag{7.36}$$

since $e^{j2\pi k} = 1$ for any value of k.
Therefore

$$A_{N-n}^* = \sum_{k=0}^{N-1} (x_k - jy_k)e^{-j2\pi kn/N}. \tag{7.37}$$

Where A^* represents the complex conjugate of A.
  Adding equations (7.35) and (7.37) gives

$$A_n + A_{N-n}^* = 2 \sum_{k=0}^{N-1} x_k e^{-j2\pi kn/N} = 2X_n$$

$$\therefore \ X_n = \frac{1}{2}(A_n + A_{N-n}^*) \tag{7.38}$$

Similarly, subtracting equation (7.37) from equation (7.35) gives

$$A_n - A_{N-n}^* = 2j \sum_{k=0}^{N-1} y_k e^{-j2\pi kn/N}$$

$$\therefore \ Y_n = \frac{1}{2j}(A_n - A_{N-n}^*). \tag{7.39}$$

  Therefore to compute the frequency spectra of each signal over the positive frequency range d.c. to $f_s/2$ we use the result of the complex DFT as follows.

A_n is given directly as the first $N/2 - 1$ computed points
A_{N-n}^* is the complex conjugate of the computed points from $N/2$ to $N-1$
  (see section 7.4.2)

  Note also that the computed coefficients are themselves complex

i.e.
$$A_n = A_{nR} + jA_{nI}$$

$$A_{N-n} = A_{(N-n)R} + jA_{(N-n)I}$$

Hence

$$X_n = \frac{1}{2}(A_{nR} + jA_{nI} + A_{(N-n)R} - jA_{(N-n)I}) \tag{7.40}$$

and

$$Y_n = \frac{1}{2j}(A_{nR} + jA_{nI} - A_{(N-n)R} + jA_{(N-n)I})$$

or $\hspace{12cm}$ (7.41)

$$Y_n = \frac{1}{2}(-jA_{nR} + A_{nI} + jA_{(N-n)R} + A_{(N-n)I}).$$

  Equations (7.40) and (7.41) provide the algorithm to compute the spectra of two real time series at one go. The only penalty is that the spectra of the individual signals must be unscrambled after the DFT of the combined signals has been computed, but this is computationally more efficient than applying the DFT to each of the signals in turn.

7.8   Improving Frequency Resolution

It has been noted that the spectra obtained from application of the DFT have a frequency range 0 to $f_s/2$, with resolution f_s/N. This is known as base-band analysis. One obvious way of increasing the frequency resolution is to increase N, the number of data points. However, in some circumstances the data may be of limited duration, nevertheless the number of points can still be augmented by appending values of zero magnitude to the time series.

$$a_p = x_k + y_m$$

$$p = 0 \rightarrow 1023 \quad k = 0 \rightarrow 511 \quad m = 512 \rightarrow 1023; \quad N = 512$$

where $y_m = 0$.

The consequence of this will be increased frequency resolution, as required, (by a factor of 2 in this case), but the amplitude of the spectral coefficients will be incorrect due to the initial scaling of the time series (by $1/P$ in the above example) and because the augmenting series is of zero magnitude and makes no contribution to the amplitude of the spectrum. The correction factor to be applied to the computed spectrum is therefore

$$\frac{P}{N} = \frac{k+m}{N}.$$

An alternative, and more flexible, approach to increasing the resolution of the analysis is to implement what has become known as 'ZOOM' analysis.

Zoom analysis yields a spectrum over a selected frequency range which lies within the normal base-band range. This type of facility is available on most digital signal processing instrumentation and whilst the details of the implementation are quite complex the basic principles of the technique are relatively straightforward.

If the DFT always yields a base-band result then the signal must be frequency shifted so that the centre frequency within the new range of interest, f_c, is at the origin, (i.e. zero frequency). This is achieved by invoking property (vii) of the DFT (see 4.5), i.e. the time series x_k is multiplied by the complex exponential $e^{j2\pi f_c t}$(or $e^{j2\pi k i/N}$) to yield a complex time series. The lower frequency of the range, f_L, now lies in the 'negative' half of the spectrum, with the upper frequency, f_u, in the positive half of the spectrum.

However, frequency shifting the signal results in an aliased spectrum, therefore at the heart of the procedure lies a very efficient low pass digital filter with good out of band rejection. The efficiency of the filter is improved by noting that the time data is over-sampled for the new frequency range ($f_u < f_s/2$), therefore if the resolution is to be improved by a factor m then only every mth complex data point requires filtering. This effectively re-samples the data at a lower sample frequency. However, we now require a record of mT seconds, if T was the initial observation time, in order to retain as many frequency-shifted and filtered data points as in the original baseband analysis. The DFT is then implemented in the usual way (via the FFT) on the modified time data. The improvements in resolution using this type of analysis can be from several Hz to mHz.

An application of the zoom technique is discussed in Chapter 9, where the transfer function of a vibratory system is under investigation.

Exercises 7

1 Plot the function $e^{-j2\pi nk/N}$, as k takes the values from 0 to $N-1$, with $n=1$ on an argand diagram i.e. real v imaginary. Use this result to explain:

(i) Why the second half of the DFT$(n=N/2+1 \rightarrow N-1)$ is the complex conjugate of the first half of the DFT.

(ii) Why the spectrum computed using the DFT is periodic, i.e. when $n=N+n$.

2 The signal from an accelerometer is to be processed digitally to determine the power spectral density of the signal. The frequency content of the signal is such that the analysis is required in the frequency range 0 to 250 Hz. Specify a suitable cut-off frequency for an anti-aliasing filter, and hence determine the sample frequency and the computed frequency range of the analysis.

How many points are required in the transform to give a frequency resolution better than 1 Hz?

If there is only a limited amount of data available for analysis then the analysis can be performed on actual data plus extra terms of zero amplitude. What will be the effect on the amplitude of the computed spectral coefficients?

3 Plot the real and imaginary parts of the discrete spectra for each of the analogue signals

$$x(t)=2\sin 8\pi t$$
$$y(t)=4\cos 6\pi t.$$

The number of samples $N=128$ and the sample rate (f_s) is 64 samples/second. Plot the positive and negative portions of the frequency spectra in each case.

This problem can be solved from first principles, refer to Chapter 4; the results can be checked using a computer program.

4 Using the results from the previous question plot the real and imaginary parts of the spectra for the time functions given below

(i) $x_1(t)=\sin 8\pi t+2\cos 6\pi t$; $N=100$; $f_s=50$ samples/s

(ii) $x_2(t)=4+3\sin 12\pi t+\cos 24\pi t$; $N=64$, $f_s=128$ samples/s

(iii) $x_3(t)=\sin 8\pi t+\sin 13.5\pi t+\cos 16\pi t$; $N=128$; $f_s=32$ samples/s.

5 Given the real and imaginary parts of a frequency spectrum (Fig. 7.35).

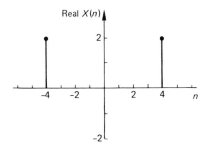
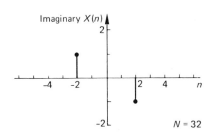

Fig. 7.35

Determine the time function that would produce such a spectrum given that the sample frequency was 32 samples/second.

6   A square wave of peak amplitude ± 4 units, period 0.1 s is sampled. The sample rate is 32 samples/s and the spectrum of the signal is based on 32 of the samples. Plot the modulus of the resulting spectrum, positive frequency range only and explain the reason for each of the spectral points.

7   A time sequence x_k is given by

$$x_k = \left(0, \frac{1}{\sqrt{2}}, 1, \frac{1}{\sqrt{2}}, 0, -\frac{1}{\sqrt{2}}, -1, -\frac{1}{\sqrt{2}} \right) \qquad k = 0, 1, \ldots$$

Re-arrange the sequence into a form suitable for the implementation of the FFT. Implement the FFT by hand calculation using the diagram in the text to lead you through the stages of the calculation.

8
Digital Filtering

8.1 Introduction

Any book concerned with signal processing must devote some space to the important subject of digital filtering. As we have mentioned several times already, the relatively low cost of computer hardware has opened up the possibility of replacing many analogue processes with their digital (or discrete) equivalent. It is in this context that we will briefly explore the fundamental concepts involved and introduce the jargon of digital filtering.

This chapter should be considered as an introduction to digital filtering. There are many excellent texts on this subject and those wishing to pursue the subject further should refer to them[9, 10, 11]. Our aim in this chapter is to show the relevance of many of the principles and techniques of signal processing discussed so far to the filtering of signals and to develop the idea of digital filtering from some basic analogue filters.

8.2 Basic Definitions

One of the most basic forms of filter, the low pass filter, can be realised by the simple resistive–capacitive network shown in Fig. 8.1.

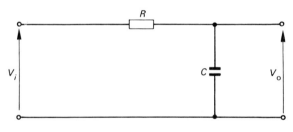

Fig. 8.1

If we apply Kirchhoff's voltage law to this circuit a first order differential equation can be devised which will relate the output voltage V_o to the input voltage V_i

$$RC\frac{dV_o}{dt} + V_o = V_i. \tag{8.1}$$

The solution of this differential equation is of an exponential form and the response (V_o) is shown in Fig. 8.2 for a step change in input voltage V_i.

When $t = RC$, $V_o \simeq 63\%$ of the final value and the time RC is defined as the time constant for the system.

The time solution to the differential equation does not appear to indicate that such a circuit would be of any practical benefit as a filter. However, if we now look at the frequency

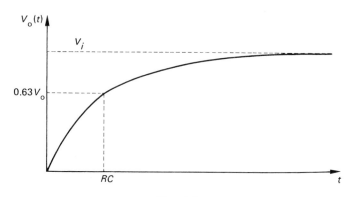

Fig. 8.2

characteristics of the system we see that such a circuit is in fact a low pass filter. In order to obtain the frequency characteristics we must first transform the differential equation (8.1) by application of the Laplace transform

$$(RCs+1)V_o(s)=V_i(s). \tag{8.2}$$

Since we are only interested in the frequency domain the substitution

$$s=j\omega$$

can be made (see Chapter 6) so that

$$(RCj\omega+1)V_o(j\omega)=V_i(j\omega) \tag{8.3}$$

or

$$\frac{V_o(j\omega)}{V_i(j\omega)}=\frac{1}{RCj\omega+1}. \tag{8.4}$$

This complex function can be written in terms of an amplitude and a phase.

i.e.

$$z=x+jy=Re^{j\theta}$$

where $|z|=R=\sqrt{x^2+y^2}$ and $\underline{/z}=\theta=\tan^{-1}y/x$

Thus

$$\left|\frac{V_o(j\omega)}{V_i(j\omega)}\right|=\frac{1}{\sqrt{[(RC\omega)^2+1]}}$$ and $\underline{/\dfrac{V_o(j\omega)}{V_i(j\omega)}}=-\tan^{-1}RC\omega. \tag{8.5}$

Plotting both of these functions as ω is varied in the range $0.1/RC\rightarrow10/RC$ yields the characteristic response shown in Fig. 8.3.

From the frequency response plot it can be seen that at low frequency ($\omega\ll1/RC$) the ratio of the output voltage to the input voltage is approximately unity so that the output voltage is relatively unaffected by the network. Note, however, that as the frequency approaches $1/RC$ the phase difference between input and output increases, with the output lagging behind the input voltage. When the frequency is above $1/RC$ the amplitude ratio decreases and so the output voltage is reduced in amplitude (attenuated) in relation to the input voltage. The phase shift now approaches $-90°$ as the frequency increases.

The important parameter is thus seen to be the time constant (RC) and the frequency associated with this ($1/RC$ or $1/T$) is known as the cut-off frequency (ω_c) since this defines the range of frequencies (the pass band) which are not significantly attenuated; components beyond this frequency being attenuated. The amplitude ratio at the cut-off frequency is

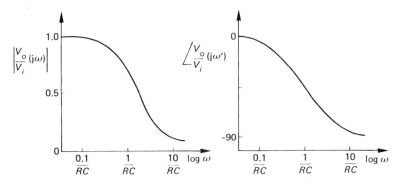

Fig. 8.3

given where

$$\left|\frac{V_o(j\omega)}{V_i(j\omega)}\right| = \frac{1}{\sqrt{2}}$$

or

$$\left|\frac{V_o(j\omega)}{V_i(j\omega)}\right|_{dB} = 20\log_{10}\left|\frac{1}{\sqrt{2}}\right|$$

which is -3 dB down compared with the low frequency components. The significance of this from the point of view of filtering is that if a signal which contains high frequency (noise) components in addition to the desired low frequency components, is passed through such a network, with appropriate values of R and C, then the high frequency components can be significantly attenuated and thus filtered; see Fig. 8.4. The filtering is of course done in the time domain and if the filter characteristic in the time domain is $g(t)$, then

$$V_o(t) = f(g(t), V_i(t)) \tag{8.6}$$

where the appropriate relationship is given by the convolution integral, as shown in Chapter 5, and $g(t)$ is the impulse response of the network

i.e.

$$V_o(t) = \int_0^t g(\tau) V_i(t-\tau)\, d\tau \tag{8.7}$$

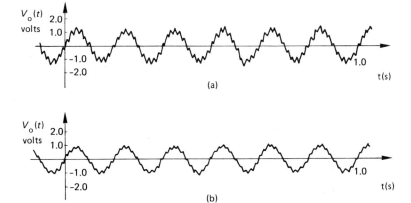

Fig. 8.4

and
$$g(t) = e^{-t/RC}. \tag{8.8}$$

Thus the filter characteristics are selected to satisfy a desired frequency response but the filter is then implemented in the time domain.

Now consider the discrete equivalent of equation (8.1) where to simplify the notation we will rewrite it as shown below:

$$T\frac{dx}{dt} + x = y \tag{8.9}$$

where x is the output from the filter
y is the input to the filter
T is the time constant for the filter (RC).

After sampling the input signal $y(t)$, a time series y_n will result with the time between samples Δ. The output signal will also be a time series if it is generated by a digital process. (Note that backward differences are used here whereas forward differences were used previously. We are following convention in using backward differences for this work, it also serves as a comparison with forward differences as used in Chapter 5).

Thus
$$y(t) \rightarrow y_n \tag{8.10}$$
$$x(t) \rightarrow x_n$$

and
$$\frac{dx}{dt} \rightarrow \frac{x_n - x_{n-1}}{\Delta}$$

since dx/dt represents the rate of change of x with respect to time (i.e. the slope of the x versus t graph at some time t_i). Substituting equation (8.10) into equation (8.9) yields

$$T\frac{(x_n - x_{n-1})}{\Delta} + x_n = y_n$$

or
$$x_n = A_0 y_n + B_1 x_{n-1} \tag{8.11}$$

where $A_0 = \dfrac{1}{1 + T/\Delta}$ and $B_1 = \dfrac{T/\Delta}{1 + T/\Delta}.$

Repeated application of equation (8.11) will yield a filtered version x_n of the time series y_n,

i.e. $x_0 = A_0 y_0$

$$x_1 = A_0 y_1 + B_1 x_0 = A_0 y_1 + B_1 A_0 y_0 \tag{8.12}$$
$$x_2 = A_0 y_2 + B_1 x_1 = A_0 y_2 + B_1 A_0 y_1 + B_1^2 A_0 y_0$$

$$\vdots$$

$$x_n = A_0 y_n + B_1 A_0 y_{n-1} + B_1^2 A_0 y_{n-2} + \ldots + B_1^n A_0 y_0.$$

If we write $g_0 = A_0$

$$g_1 = B_1 A_0 \tag{8.13}$$
$$g_2 = B_1^2 A_0$$

$$\vdots$$

$$g_n = B_1^n A_0$$

then $\qquad x_n = g_0 y_n + g_1 y_{n-1} + g_2 y_{n-2} + \ldots + g_n y_0$ (8.14)

or $\qquad x_n = \sum_{k=0}^{n} g_k y_{n-k}.$ (8.15)

Comparing equations (8.7) and (8.15) we see that the implementation of equation (8.11) is in fact the discrete equivalent of the convolution integral which was previously discussed in Chapter 5.

Before proceeding let us consider how such a filter might be of practical use (without using any further knowledge about the design of digital filters). Consider a pure sinusoid contaminated by 50 Hz mains hum so that the observed signal is as shown in Fig. 8.5.

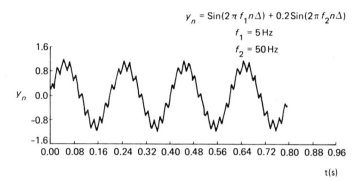

$$y_n = \text{Sin}(2\pi f_1 n\Delta) + 0.2\,\text{Sin}(2\pi f_2 n\Delta)$$
$$f_1 = 5\,\text{Hz}$$
$$f_2 = 50\,\text{Hz}$$

Fig. 8.5

Assume that the desired signal has a frequency of below 20 Hz then we can select the filter cut-off frequency to be 20 Hz and so the time constant T of the filter is given by

$$T = \frac{1}{2\pi \cdot 20} = 7.95 \text{ ms.}$$

We must now select a suitable sample interval. The choice of sample interval (or sample rate) has already been discussed at some length in Chapters 1 and 7 so that, to represent the noisy signal the sample rate must be at least 100 Hz and for good representation at least 200 Hz. For a rate of 250 Hz the sample interval Δ is given by

$$\Delta = \frac{1}{250} = 4 \text{ ms.}$$

The coefficients of the filter equation (8.11) are now given by

$$A_0 = 0.3345 \qquad \text{and} \qquad B_1 = 0.6655$$

so that the difference equation defining the filter becomes

$$x_n = 0.3345 y_n + 0.6655 x_{n-1}.$$ (8.16)

Applying the equation to the sampled time series yields the filtered time series shown in Fig. 8.6. Now consider the effect of increasing the filter cut-off frequency ω_c, Fig. 8.7(a), and reducing the sample rate, Fig. 8.7(b), respectively. We see that there are two design parameters for the digital filter as both the cut-off frequency and the sample rate influence the effectiveness of the filter.

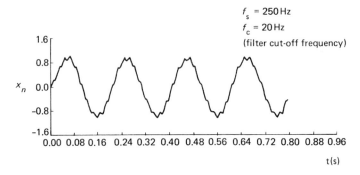

Fig. 8.6

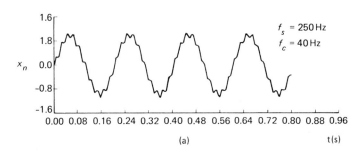

(a)

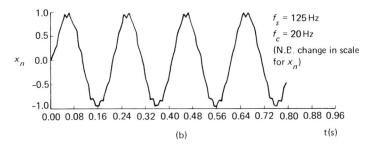

(b)

Fig. 8.7 ·

8.2.1 Comments

A simple analogue low pass filter is relatively easy to realise as an electrical network. The characteristics of the filter can be assessed from the transfer function of the filter and by plotting the frequency response curves.

An equivalent digital filter can be realised by forming the difference equation and evaluating the coefficients for a particular application. So far the filter parameters have been selected from experience since we have not investigated the frequency characteristics of the digital filter.

Both the analogue and digital filter are implemented in the time domain via convolution although in many applications this will not be immediately apparent.

8.3   The Pulse Transfer Function and Frequency Characteristics

As we have seen the digital filter can be implemented by a difference equation of the form shown below.

$$x_n = A_0 y_n + B_1 x_{n-1} \qquad (8.17)$$

where x_n and y_n are the present values.

x_{n-1} is the value of x at the previous sample interval, i.e. one time delay previously. Alternatively this can be written as

$$x(z) = A_0 y(z) + B_1 z^{-1} x(z) \qquad (8.18)$$

where z^{-1} represents one time delay.
Therefore

$$x(z)(1 - B_1 z^{-1}) = A_0 y(z) \qquad (8.19)$$

thus

$$\frac{x(z)}{y(z)} = \frac{A_0}{1 - B_1 z^{-1}}. \qquad (8.20)$$

We have now derived a transfer function to represent the filter by assuming that the operator z is a linear algebraic operator. The operator is the z-transform which was introduced in Chapter 5; the properties of the z-transform and its relationship with the Laplace transform are given in Appendix B, (i.e. $z = e^{s\Delta}$).

The transfer function given by equation (8.20) is known as the pulse transfer function from which we can now obtain the frequency characteristics of the digital filter. Remembering that we used the substitution ($s = j\omega$) for the analogue filter we can now use a similar substitution

$$z = e^{j\omega\Delta}(z^{-1} = e^{-j\omega\Delta})$$

so that equation (8.20) becomes

$$\frac{x}{y}(j\omega) = \frac{A_0}{1 - B_1 e^{-j\omega\Delta}} = G(z) \text{ where } \quad z = e^{j\omega\Delta} \qquad (8.21)$$

The frequency characteristics of the filter specified in section 8.1 can now be examined by using the appropriate values of A_0, B_1 and Δ in equation (8.21)

$$G(z) = \frac{0.3345}{1 - 0.6655\, e^{-j(0.004\,\omega)}};$$

And plotting this function over the frequency range 1–100 Hz yields the amplitude characteristics shown in Fig. 8.8. Note that at the expected break frequency (20 Hz) the amplitude characteristic is almost -5 dB. Therefore the analogue break frequency, (i.e. -3 dB down frequency), does not translate across to the digital filter. There is therefore a relationship between analogue and digital frequencies and we will derive this in section 8.5.

Alternative methods of deriving the digital transfer function are:

(i)   partial fraction expansion of the standard (Laplace transformed) transfer function and then transformation to the 'z' domain;

(ii)   use of the bilinear transformation between 's' and 'z'.

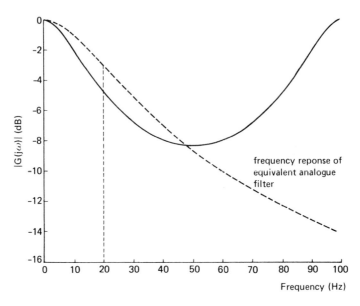

Fig. 8.8

To demonstrate both of these methods, consider again the *RC* network and rewriting equation (8.2) so that *x* represents the filter output and *y* the input

$$(RCs+1)X(s) = Y(s). \tag{8.22}$$

The transfer function is then

$$\frac{X(s)}{Y(s)} = \frac{1}{RCs+1} = G(s). \tag{8.23}$$

Example 8.1

The partial fraction expansion method is of course straightforward for the first order system defined by equation (8.23) since the denominator is in reduced form. The *z* transform of this transfer function can now be obtained by inspection from Table A.2 i.e.

$$\frac{b}{s+a} \Leftrightarrow \frac{b}{1-e^{-a\Delta}.z^{-1}}.$$

In our case

$$a = 1/RC$$
$$b = 1/RC$$

so that

$$G(z) = \frac{\Delta/RC}{1-e^{-\Delta/RC}.z^{-1}} = \frac{x(z)}{y(z)}. \tag{8.24}$$

After introducing the sample interval as a scale factor (see Chapter 5), we get a result which is consistent with the impulse invariance method which is discussed in the literature on digital filtering.[9]

Rearranging equation (8.24) we get

$$(1-e^{-\Delta/RC}.z^{-1})x(z) = \Delta/RC.y(z) \tag{8.25}$$

and rewriting in the notation of equation (8.19) and setting $RC = T$ yields

$$x(z) - e^{-\Delta/T} z^{-1} x(z) = \Delta/T y(z)$$

or

$$x(z) = \Delta/T y(z) + e^{-\Delta/T} z^{-1} x(z)$$

$$x(z) = A_0 y(z) + B_1 z^{-1} x(z). \tag{8.26}$$

In this case

$$A_0 = \Delta/T$$
$$B_1 = e^{-\Delta/T}.$$

This gives the difference equation

$$x_i = A_0 y_i + B_i x_{i+1}$$

Example 8.2

The bilinear transformation between the s and z domains is given by[9]

$$s = \frac{2}{\Delta} \cdot \frac{1 - z^{-1}}{1 + z^{-1}}. \tag{8.27}$$

Substituting for s in the transfer function defined by equation (8.23) (again let $RC = T$) we obtain

$$G(z) = \frac{1}{\dfrac{2T}{\Delta} \dfrac{1 - z^{-1}}{1 + z^{-1}} + 1} = \frac{1 + z^{-1}}{\dfrac{2T}{\Delta}(1 - z^{-1}) + (1 + z^{-1})}$$

$$G(z) = \frac{\dfrac{1}{2T/\Delta + 1} + \dfrac{1}{2T/\Delta + 1} \cdot z^{-1}}{1 - \dfrac{\left(\dfrac{2T}{\Delta} - 1\right) z^{-1}}{\left(\dfrac{2T}{\Delta} + 1\right)}} = \frac{A_0 + A_1 z^{-1}}{1 - B_1 z^{-1}} \tag{8.28}$$

where

$$A_0 = A_1 = \frac{1}{\dfrac{2T}{\Delta} + 1}$$

$$B_1 = \frac{2T/\Delta - 1}{2T/\Delta + 1}.$$

Rearranging equation (8.28) in terms of the time series $x(z)$ and $y(z)$

$$x(z)(1 - B_1 z^{-1}) = y(z)(A_0 + A_1 z^{-1})$$

or

$$x(z) = A_0 y(z) + A_1 z^{-1} y(z) + B_1 z^{-1} x(z). \tag{8.29}$$

In this case the difference equation is

$$x_i = A_0 y_i + A_1 y_{i-1} B_1 x_{i-1}.$$

We now have three versions of the digital '*RC*' filter, so we can compare coefficients in general and calculate values for the filter selected in section 8.2. These are shown in Table 8.1

We can now compare the characteristics of three versions in both the time and frequency domains, see Figs. 8.9 and 8.10.

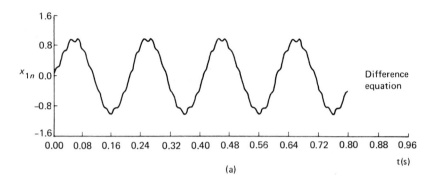

(a)

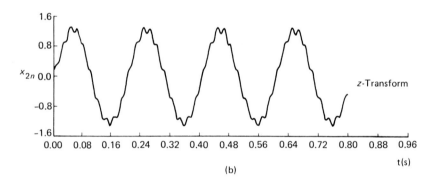

(b)

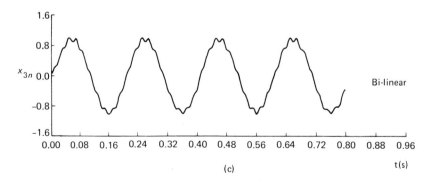

(c)

Fig. 8.9

Table 8.1

Method of derivation	Coefficient					
	B_1	A_0	A_1	B_1	A_0	A_1
				(Numerical values)		
Difference equation	$\dfrac{T/\Delta}{(1+T/\Delta)}$	$\dfrac{1}{(1+T/\Delta)}$	—	0.6655	0.3345	—
Standard z transform	$e^{-\Delta/T}$	Δ/T	—	0.6049	0.5026	—
Bilinear transformation	$\dfrac{(2T/\Delta-1)}{(2T/\Delta+1)}$	$\dfrac{1}{(2T/\Delta+1)}$	$\dfrac{1}{(2T/\Delta+1)}$	0.5983	0.2008	0.2008

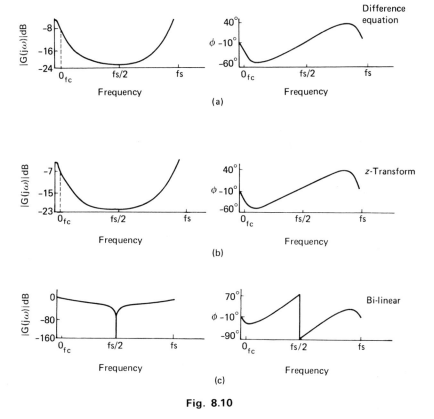

Fig. 8.10

At this stage it should be pointed out that filters, other than the simple first order low pass type discussed, can be synthesised from their analogue equivalents. The equivalent high pass filter, for example, is obtained by changing the sign in the denominator of the digital transfer function (equation 8.20).

$$G(z)=\frac{A_0}{1+B_1z^{-1}}.\tag{8.30}$$

The design of digital filters from the analogue equivalent is an accepted approach but the implementation of such filters should be done with some care particularly when the difference equation approach is used. The reason for this is that for higher order ($>$3rd order) difference equations the numerical accuracy deteriorates very rapidly. To overcome this problem it is convenient to factor the equations so that the desired transfer function[9] is given by

$$G(z) = G_1(z) \times G_2(z) \times G_3(z) \times \ldots G_m(z). \tag{8.31}$$

An interesting application of digital filtering is the repeated filtering of a signal to reduce the amplitude of unwanted noise. Consider the sequence of results shown in Fig. 8.11 where the original signal has been filtered repeatedly using the same filter coefficients. The

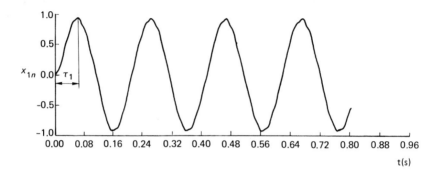

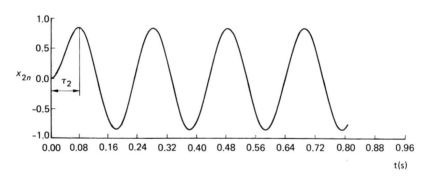

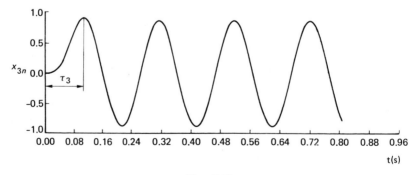

Fig. 8.11

amplitude of the desired sinusoid has been corrected at each stage of filtering but the phase lag has been left in the uncorrected form. $(\tau_1 < \tau_2 < \tau_3)$

The effect of repeated filtering is the same as passing the signal through a higher order filter but, again, beware of numerical rounding errors.

8.4   The Impulse Response of Digital Filters—Filter Classification

We have already discussed the importance of the impulse response of a system in the context of linear system theory in Chapter 5, and also the relevance of the impulse response to filters in section 8.1. Now consider the impulse response of a digital filter.

We have previously implied that a digital filter has an impulse response, e.g. equation 8.15, where g_n defines a system characteristic. Returning to the general form of the difference equation for a first order low pass filter

$$x_n = A_0 y_n + B_1 x_{n-1} \tag{8.32}$$

then the impulse response is given by the values of x_n where the input y_n takes the form

$$y_0 = 1$$

$$y_i = 0 \quad \text{for } i = 1, n \qquad \text{(cf. equation (1.9)).} \tag{8.33}$$

Thus

$$x_0 = A_0$$

$$x_1 = B_1 x_0 = B_1 A_0$$

$$x_2 = B_1 x_1 = B_1^2 A_0$$

$$x_3 = B_1 x_2 = B_1^3 A_0$$

$$\vdots$$

$$x_n = B_1 x_{n-1} = B_1^n A_0 \tag{8.34}$$

and hence g_n defined by equation (8.13) is identical to equation (8.34) and so g_n is the impulse response of the system. Alternatively, consider the result obtained by dividing the denominator of the digital transfer function into the numerator.

$$G(z) = \frac{A_0}{1 - B_1 z^{-1}} \tag{8.35}$$

$$G(z) = A_0 + B_1 A_0 z^{-1} + B_1^2 A_0 z^{-2} + B_1^3 A_0 z^{-3} + \ldots + B_1^n A_0 z^{-n}. \tag{8.36}$$

The coefficients of the infinite series which results from the long division are equal to those of the time series in equations (8.34). Hence, from the analogy with the Laplace transform, the impulse response obtained by long division is equivalent to the inverse transform.

One reason for investigating the impulse response of a digital filter is that the form of the response provides a means of classifying the type of filter. There are two types of filter response, the infinite impulse response (IIR), typified by equation (8.36) and the finite impulse response (FIR). This, by definition, has a finite number of terms in the impulse response, e.g.

$$G(z) = 1 + A_1 z^{-1} + 0 z^{-2} + 0 z^{-3} + \ldots + 0 z^{-n} \tag{8.37}$$

where all coefficients beyond the second term are zero in this example.

Digital filters are also classified by the form of the difference equation and, again, these fall into two distinct types, recursive and non-recursive. The recursive filter has already been discussed and has the form of equation (8.29) where the output x_n, at some particular time, depends not only on the input but also on previous values of the output. In contrast the output of a non-recursive filter depends only on the input and takes the form shown in equation (8.38).

$$x_n = y_n + A_1 y_{n-1} \qquad (8.38)$$

In general both FIR and IIR filters can be implemented as recursive or non-recursive filters. However, the simplest FIR filter takes the non-recursive form and the simplest IIR the recursive form.

8.5   Digital Filtering Via the FFT (Fast Convolution)

The relationship between convolution in both the time and frequency domains has been discussed at length in previous chapters. We know, therefore, that the convolution summation, which is effectively the means by which a digital filter is implemented, can be replaced by a multiplicative operation in the frequency domain. Since multiplication of terms is significantly faster, from a computational viewpoint than convolution, it is worth investigating the possibility of using the FFT to implement the filtering procedure.

The filtering procedure is shown diagrammatically in Fig. 8.12 as both convolution in time and multiplication in frequency.

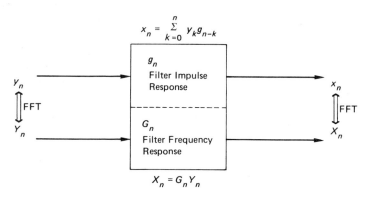

Fig. 8.12

From our knowledge so far, the steps to follow in order to utilise the FFT appear to be:

(i)   transform the input time series using the FFT algorithm;
(ii)   multiply the transformed time series by the filter frequency response;
(iii) form the inverse transform of the product to obtain the desired filtered time series.

Unfortunately, if this procedure is followed the results obtained will be in error. The reason for this is once again due to the periodicity which the DFT forces on both the time sequence and the resulting spectrum (see section 7.4.1).

To demonstrate the effect of periodic time series on convolution consider the time series y_n and the filter impulse response g_n shown in Fig. 8.13.

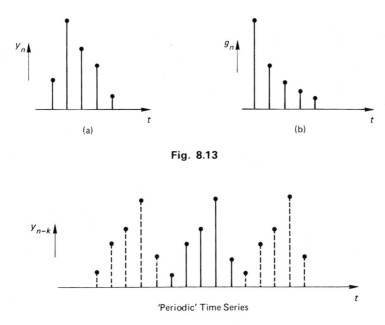

Fig. 8.13

Fig. 8.14

'Periodic' Time Series

The convolution summation (equation 8.15) requires that the modified time series y_{n-k} be formed (as shown in Chapter 5) so that this time series appears as shown in Fig. 8.14. The effect of the assumed periodicity is shown in dotted lines.

Quite clearly the result of convolving the time series shown in Fig. 8.14 with the filter impulse response will be significantly different from convolution using the original time series. To illustrate this, consider the relative positions of the two versions of y_{n-k} after the 6th time shift, see Fig. 8.15.

The distortion due to the periodicity of the time series is known as 'wrap around' and leads to 'circular' convolution. To overcome this effect it is necessary to augment both the

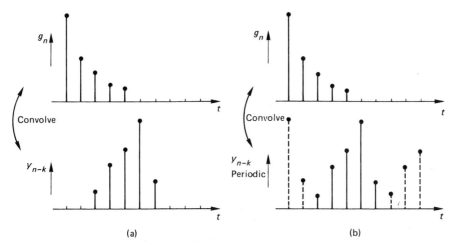

Fig. 8.15

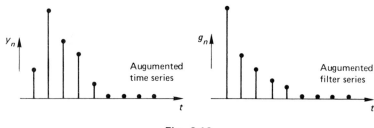

Fig. 8.16

time series and the filter response with zeros in order to extend the period of the time series and filter response outside the range of the computation. The augmented time series and filter response are then as shown in Fig. 8.16.

To eliminate 'wrap around' the following condition must be satisfied:

$$N > P + Q - 1 \tag{8.39}$$

where

N is the number of points in each of the augmented time series
P is the number of data points and
Q is the number of filter points.

To utilise the FFT algorithm developed in Chapter 7, N must also satisfy the constraint that

$$N = 2^m.$$

The additional steps in the procedure to implement convolution via the FFT are as follows:

(a)   Choose the number of sample points to be transformed which eliminate 'wrap around', i.e. apply equation (8.39) and

(b)   augment both the time series and filter response with the required number of zeros. Now continue with steps (i) to (iii) as previously discussed.

A further practical problem which arises is that in general the time series to be filtered will be much longer than the impulse response of the filter. In some applications the time sequence may even be of infinite length. The solution to this problem is to decompose the time series into smaller sections as shown in Fig 8.17.

Each section is now of length L, and all other points outside the range of the section are set to zero. The original time series is now the sum of each of the sectioned sequences.

If the impulse response of the filter is of length Q where $Q < L$, then to avoid circular convolution the combined length of both the filter and each section of the signal must be $(L + Q - 1)$. In addition, the sections of the signal and filter response must be augmented with zeros. The FFT can then be used to implement indirect convolution as described in the previous section to produce filtered sections of the data. The complete filtered signal is then reconstructed by adding together each of the filtered sections in the same way that the orignal signal was reconstructed from the sub-sections. This method is known as the overlap–add method since the filtered sections are overlapped, due to the augmenting procedure, and then added as shown in Fig. 8.18.

8.6   The Design of Digital Filters

As we have already seen the impulse function can be derived directly from analogue transfer functions expressed in terms of the Laplace transform. This approach to the design of digital

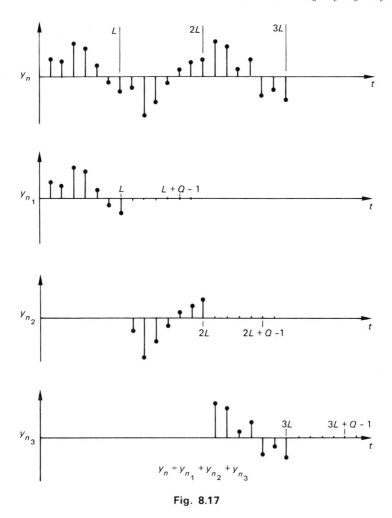

Fig. 8.17

filters will therefore be examined further. An alternative approach is to design a filter which takes advantage of the fast convolution technique previously discussed. Both of these design techniques will be illustrated by means of examples. Once again, the problem is to design suitable filters to smooth the sine wave shown in Fig. 8.5.

Example 8.3

One class of analogue filters is known as a Butterworth filter. The modulus squared of the frequency response function for this type of filter takes the characteristic form

$$|G(j\omega)|^2 = \frac{1}{1+(\omega/\omega_c)^{2n}} \left(\text{or } |G(f)|^2 = \frac{1}{1+(f/f_c)^{2n}} \right) \tag{8.40}$$

where f_c is the 3 dB down, cut-off frequency, and n the order of the filter transfer function, i.e. the number of system poles. The design specification for the digital filter used in the previous example, is that the cut-off frequency is 20 Hz and the amplitude must be at least

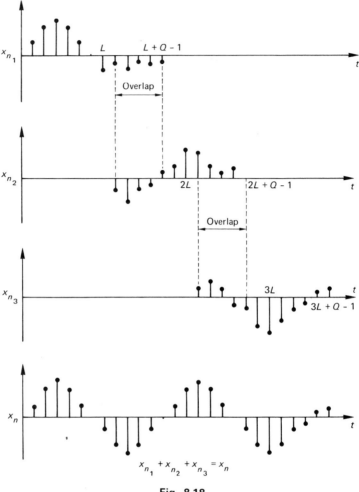

Fig. 8.18

20 dB down at 50 Hz, the known contamination frequency. The sample rate is 250 Hz, which was used in previous examples.

Before we can proceed with the design, however, we must convert the frequencies, expressed in terms of digital requirement, to the equivalent analogue frequencies. The necessity to do this was established in section 8.2. This can be done using a transformation derived from the bilinear z transform.

$$s = \frac{2}{\Delta} \times \frac{1-z^{-1}}{1+z^{-1}}. \tag{8.41}$$

Expressing this equation in terms of frequency we substitute $s = j\omega_a$ and $z = e^{-j\omega_d \Delta}$

then
$$j\omega_a = \frac{2}{\Delta} \times \frac{1-e^{-j\omega_d \Delta}}{1+e^{-j\omega_d \Delta}} \tag{8.42}$$

where ω_d is the 'digital' frequency,

ω_a is the equivalent analogue frequency.

Now remember that $\cos \omega t$ and $\sin \omega t$ can be expressed in terms of complex exponentials so that

$$j\sin\left(\frac{\omega_d \Delta}{2}\right) = \tfrac{1}{2}(e^{j\omega_d \Delta/2} - e^{-j\omega_d \Delta/2})$$

and

$$\cos\left(\frac{\omega_d \Delta}{2}\right) = \tfrac{1}{2}(e^{j\omega_d \Delta/2} + e^{-j\omega_d \Delta/2}).$$

Thus

$$\frac{j\sin(\omega_d \Delta/2)}{\cos(\omega_d \Delta/2)} = \frac{1 - e^{-j\omega_d \Delta}}{1 + e^{-j\omega_d \Delta}}. \tag{8.43}$$

Substituting this result into equation (8.42) yields the required frequency transformation, i.e.

$$\omega_a = \frac{2}{\Delta} \times \frac{\sin(\omega_d \Delta/2)}{\cos(\omega_d \Delta/2)} = \frac{2}{\Delta}\tan\left(\frac{\omega_d \Delta}{2}\right). \tag{8.44}$$

Recalling that Δ, the sample interval, is related to the sample frequency f_s (expressed in Hz)

$$\omega_a = \frac{2}{\Delta}\tan\left(\frac{\pi f_d}{f_s}\right)$$

where f_d is the digital frequency in Hz, and introducing f_a the analogue frequency in Hz we have

$$f_a = \frac{1}{\pi \Delta}\tan\left(\frac{\pi f_d}{f_s}\right). \tag{8.45}$$

We can now make use of this relationship to convert the digital specification for the filter to an analogue specification.

The 'digital' cut-off frequency is specified to be 20 Hz (f_{cd})

$$\therefore f_{ca} = \frac{1}{\pi \Delta}\tan\left(\frac{20\pi}{250}\right) = \frac{0.257}{\pi \Delta}.$$

The other frequency specified is the -20 dB point at 50 Hz

$$\therefore f_a = \frac{1}{\pi \Delta}\tan\left(\frac{50\pi}{250}\right) = \frac{0.727}{\pi \Delta}$$

$$\therefore \frac{f_a}{f_{ca}} = 2.83 \left(= \frac{\omega_a}{\omega_{ca}}\right). \tag{8.46}$$

The Butterworth filter can now be defined in terms of analogue frequencies

$$|G(f)|^2 = \frac{1}{1 + (2.83)^{2n}} = 10^{-2} \; (-20 \text{ dB and } 50 \text{ Hz})$$

$$\therefore \; 1 + (2.83)^{2n} \geqslant 10^2$$

$$n \geqslant 2.2. \tag{8.47}$$

Since n must be an integer (it defines the number of poles) the required value is $n = 3$.

This means that the filter can be realised by a system, the transfer function of which has three poles equally spaced around the origin of the s-plane as shown in Fig. 8.19. Note that the poles must all lie in the left half plane in order to satisfy the conditions for stability.

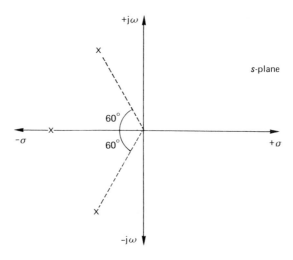

Fig. 8.19

The transfer function of the filter is given by

$$G(s) = \frac{\omega_{ca}^3}{(s + \omega_{ca})(s^2 + \omega_{ca}s + \omega_{ca}^2)}. \tag{8.48}$$

Thus the poles are:

$$s_1 = -\omega_{ca} \qquad = -0.257$$

$$s_2 = -\omega_{ca}(\tfrac{1}{2} + j\sqrt{3/2}) = -0.257\left(\frac{1}{2} + j\frac{\sqrt{3}}{2}\right)$$

$$s_3 = -\omega_{ca}(\tfrac{1}{2} - j\sqrt{3/2}) = -0.257\left(\frac{1}{2} - j\frac{\sqrt{3}}{2}\right)$$

Note that the scale factor has been omitted from the equivalent analogue frequency. This is because the factor cancels out within the transfer function.

The transfer function is now given by

$$G(s) = \frac{\omega_{ca}^3}{(s + s_1)(s + s_2)(s + s_3)}.$$

Substituting for s_1, s_2, s_3 and evaluating the denominator as a polynomial we have

$$G(s) = \frac{0.017}{s^3 + 0.514s^2 + 0.132S + 0.017}. \tag{8.49}$$

To obtain the impulse function we can derive the z-transform of the transfer function by applying the bilinear transformation (equation 8.27). Again, the factor $(2/\Delta)$ can be dropped and so the pulse transfer function becomes

$$G(z) = \frac{0.017}{\left(\dfrac{1 - z^{-1}}{1 + z^{-1}}\right)^3 + 0.514\left(\dfrac{1 - z^{-1}}{1 + z^{-1}}\right)^2 + 0.132\left(\dfrac{1 - z^{-1}}{1 + z^{-1}}\right) + 0.017}. \tag{8.50}$$

Expanding, rearranging and collecting terms together yields

$$G(z) = \frac{0.0102(1 + 3z^{-1} + 3z^{-2} + z^{-3})}{1 - 2.003\,z^{-1} + 1.446z^{-2} - 0.361z^{-3}} = \frac{y(z)}{x(z)}. \tag{8.51}$$

The required difference equation is now obtained by substituting $z^{-1}x(z) = x_{n-1}, z^{-2}x(z) = x_{n-2}$ etc.

$$y_n = 2.003\,y_{n-1} - 1.446\,y_{n-2} + 0.361\,y_{n-3}$$
$$+ 0.0102(x_n + 3x_{n-1} + 3x_{n-2} + x_{n-3}). \tag{8.52}$$

The final difference equation can be programmed relatively easily to filter data. The effectiveness of the filter can be judged from Fig. 8.20 where, once again, the sine plus 50 Hz noise signal (Fig. 8.5) has been used as the test signal. Remember in previous tests the filter cut off frequency was also 20 Hz and sample rate 250 Hz.

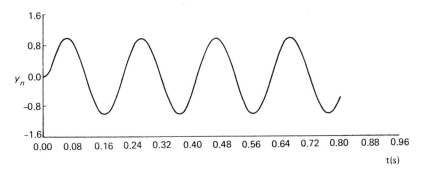

Fig. 8.20

Example 8.4

The filters which have been designed so far have all been implemented in the time domain and the final difference equations have all been recursive. This type of equation results from the design of the digital filter from an equivalent analogue filter. An alternative approach is to design the filter from an idealised frequency response characteristic and then obtain the impulse function associated with this frequency characteristic. By way of example consider an ideal low pass filter with the amplitude characteristics shown in Fig. 8.21. The filter has a

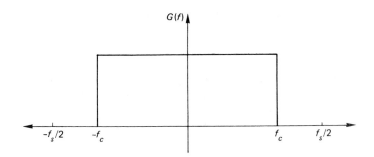

Fig. 8.21

cut-off frequency f_c and the sample frequency is f_s. The impulse response of this function is given by the inverse Fourier transform of the frequency function which can be deduced from the direct Fourier transform of the single square function derived in section 4.4.

$$g(t) = 2f_c \cdot \frac{\sin(2\pi f_c t)}{2\pi f_c t}.$$

The time series for the impulse response is therefore given by (noting that the function is normalised by multiplying by Δ).

$$g_n = 2f_c \Delta \frac{\sin(n 2\pi f_c \Delta)}{n 2\pi f_c \Delta}. \tag{8.53}$$

In order to illustrate what the time series looks like substitute $f_s = 200$ Hz, (i.e. $\Delta = 200^{-1}$) and $f_c = f_s/10$,

i.e.
$$f_c \Delta = \frac{f_s}{10} \cdot \frac{1}{f_s} = \frac{1}{10}$$

$$g_n = \frac{1}{5} \cdot \frac{\sin(n\pi/5)}{n\pi/5}. \tag{8.54}$$

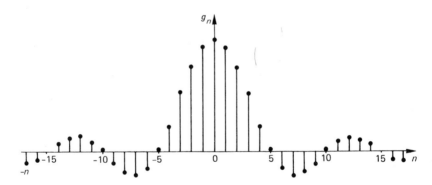

Fig. 8.22

Plotting this yields the discrete impulse response shown in Fig. 8.22. This function clearly has two disadvantages:

1)   It is of infinite duration although the amplitude tends to zero as $n \to \infty$.
2)   The function cannot be implemented on real signals since negative time is implied, i.e. as $n \to -\infty$.

To overcome these two problems the impulse response must be truncated and the function shifted in time so that the entire response is in positive time. The function now becomes as shown in Fig. 8.23.

The impulse sequence is now said to be 'causal' and the filter can be realised since the sequence is zero for $n < 0$.

However, two further problems are introduced. Once again we have truncated a time series and therefore it has been windowed. Also the effect of a time shift in the time domain

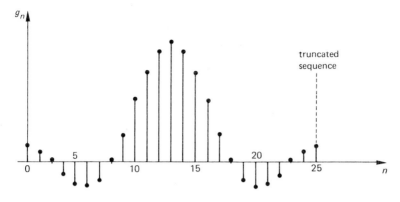

Fig. 8.23

is to cause a linear phase shift in the frequency domain

$$\text{Phase shift } \phi = m2\pi f\Delta$$

where $m =$ number of time shifts to yield a causal time sequence.

Although we started with an ideal filter response the manipulations performed on the impulse response result in a distortion of the amplitude characteristics, see Fig. 8.24. (The frequency function was obtained by applying the FFT to the impulse series augmented by zeros. This results in the improved frequency resolution shown.)

To improve the filter characteristics, window functions are used. Two window functions which have the necessary properties for digital filters are the Hamming and Kaiser functions.

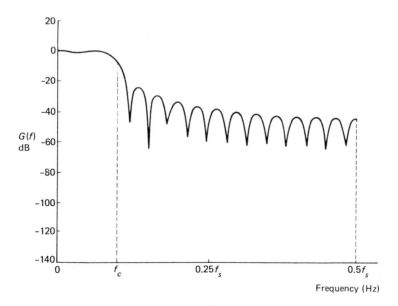

Fig. 8.24

The Hamming window is defined in continuous time by the function

$$w_H(t) = 0.54 + 0.46 \cos\left(2\pi\left(t - \frac{T}{2}\right)\bigg/ T\right) \tag{8.55}$$

where T is the total record length in seconds and t the instantaneous time. In discrete time this becomes

$$w_H(n) = 0.54 + 0.46 \cos\left(2\pi\left(n - \frac{N}{2}\right)\bigg/ N\right)$$

where N is the total number of samples and n the current sample.

The Kaiser window is defined by

$$w_K(n) = \frac{J_0(2\pi(1 - (n/N)^2)^{1/2}}{J_0(2\pi)} \tag{8.56}$$

where

$$J_0(x) = 1 + \sum_{m=1}^{M}\left[\frac{(x/2)^m}{m!}\right]^2 \tag{8.57}$$

which is a Bessel function of the first kind and zero order. The upper limit M, on the value of m is chosen so that the series converges, (e.g. $M = 100$). Applying each window to the truncated causal impulse response and then transforming to the frequency domain yields the frequency response functions shown in Fig. 8.25. From Fig. 8.25 we can clearly see the benefit of using a window function to compensate for the effects of truncation. The improved attenuation gained from the window functions, however, is at the expense of the initial roll-off rate but this will normally be acceptable and is the usual trade-off whenever windowing is used.

We are now in a position to filter the contaminated sine function. The overlap–add method was programmed and the results of the filtering algorithm are shown in Fig. 8.26.

Note that there is an initial delay in the computed filtered signal. This is due to the delay introduced in making the impulse function causal. For the filter parameters chosen here the delay is 10 samples.

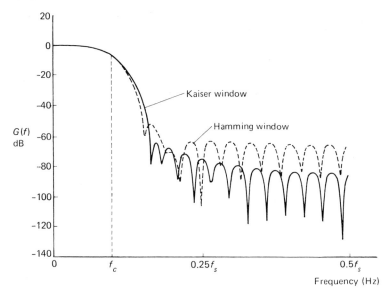

Fig. 8.25

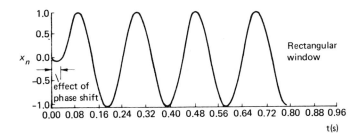

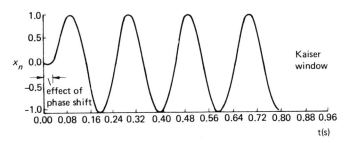

Fig. 8.26

8.7 Concluding Comments

Several different methods of 'designing' digital filters have been introduced. The principal methods which have been discussed are based on established analogue filter designs. However, digital filters can be synthesised from idealised frequency response characteristics and in theory a filter of any desired characteristic can be produced. There is, of course, the constraint of translating idealised theoretical concepts into reality where we can only handle limited amounts of data to finite resolution. This results in only an approximation to the idealised filter being attainable.

The majority of filtering problems can be solved by using one of the methods described in this chapter. Indeed, many signal conditioning problems can be overcome by using a simple first order filter which was discussed in section 8.2.

Exercises 8

1 Show that the amplitude and phase response of a filter defined by:

$$G(\omega) = \frac{1}{1 + 0.4\,e^{-j\omega\Delta}}$$

is given by

$$|G(\omega)| = \frac{0.93}{\sqrt{1 + 0.69\cos(\omega\Delta)}}; \qquad \phi = -\tan^{-1}\left(\frac{0.4\sin(\omega\Delta)}{1 + 0.4\cos(\omega\Delta)}\right).$$

2   Write the following equations in pulse transfer function form and hence determine the closed form solution for each equation when the input y_n is a unit step. The solution of difference equations is discussed in the Appendix.

$$x_n = y_n + 0.5\, x_{n-1}$$

$$x_n = y_n - 0.5\, x_{n-1}.$$

Plot the first five values of each response. Calculate the steady state value in each case and confirm the results by comparison with the appropriate time response.

Can you deduce the filtering effect that would result from application of the equations to time data?

Calculate and plot the frequency response characteristics of each filter, i.e. $|G(j\omega)|v(\omega\Delta)$ for the range of values of $(\omega\Delta) = 0.1\pi$, 0.2π, $\ldots$ π. Now comment on the function of each filter. Can you predict the form of the frequency response when $(\omega\Delta)$ takes values in the ranges $\pi \to 2\pi$, $2\pi \to 4\pi$, $4\pi \to 6\pi$ etc.

3   The transfer function for a first order analogue filter is given by;

$$G(s) = \frac{1}{0.01s + 1}.$$

Determine the cut-off frequency of the filter and compare this with the corresponding values for the digital filter when the discretisation is based on

(i)   finite differences
(ii) bi-linear transformation, and
(iii) standard z-transform

Sample rate 100 samples/second

The digital filter, based on finite differences, is used to eliminate an unwanted periodic component in the signal at 20 Hz. The component must be attenuated by at least 20 dB. How many times should the data be passed through the filter to achieve this?

4   The analogue specification for a filter is given by

$$G(s) = \frac{0.5}{s + 0.5}.$$

Use the bi-linear transformation to derive the pulse transfer function and show that the digital and analogue cut-off frequencies are related by equation 8.45. Determine the difference equation for the filter.

5   Use the Butterworth filter approach to design a digital filter to satisfy the following criteria. Cut off frequency 100 Hz; the amplitude of the filter must be at least 10 dB down at 200 Hz and the sample frequency is 1 kHz. Determine the pulse transfer function for the filter and hence the difference equation which would be used to implement the filter.

9
Case Studies

9.1 System Identification using Random Signals—Principles

We have seen that, in order to study a system and/or predict its behaviour, it is necessary to have an adequate model; this can be in the form of a differential equation or a transfer function etc. Whilst modelling is a wide subject and beyond the scope of our present studies we have indicated that one way of finding a model is to build one up from the known physical attributes of the system under study. (See for example section 5.2 and Appendix C.) Another method of finding a model is to perform special tests on the unknown system and then process the results appropriately. This latter method is known as system identification but it clearly requires the system to be a physical reality before it can be implemented. The former approach can be used to model proposed systems and is useful at the design stage whereas the latter can be used for model validation after a system has been built.

There are many ways of performing system identification and all amount to a signal processing operation, indeed whole books have been written on the subject[15], however, we limit ourselves to simple case studies for the purpose of illustration.

In section 5.13 we found that, if the input to a system is white noise, then the cross-correlation function relating system output with input is proportional to its impulse response. (Equation 5.26).

$$R_{uy}(\tau) = Ng(\tau).$$

This fact can be used as the basis of an experiment to determine the impulse response of a system whose characteristics are unknown. The unknown system is excited with white noise and the input–output cross-correlation is measured; the result usually appears in graphical form. The above result tells us that this graph also represents the required impulse response. Since an algebraic form for the impulse response is more convenient we run into the difficulty of finding a function which fits the graph and a curve fitting exercise is now necessary. However, here, our interest lies mainly in the signal processing aspects of the problem which we now take up further.

As one would expect, in reality the performance of an experiment is fraught with many difficulties and one which is ubiquitous to any is that of unwanted disturbances or noise. However, we can very easily show that, provided certain conditions hold, this cross-correlation method is immune to noise disturbance.

Refer to Fig. 9.1. Let the unknown system be subject to a random input, u, under control from a generator and assume that unwanted noise disturbances can be regarded as another input, n. The unknown impulse response $g(t)$ is to be determined by cross-correlation of output z with input u. However z will be corrupted by a component emanating from n. Let the impulse response of the system due to n be $h(t)$, the output due to n be w, and the output due to u be y. Assuming that the unwanted component w is additive we can say that

$$z = y + w$$

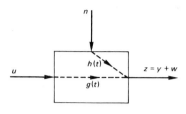

Fig. 9.1

and
$$z(t) = y(t) + w(t)$$

But
$$y(t) = \int_0^\infty g(\eta)u(t-\eta)\mathrm{d}\eta$$

and
$$w(t) = \int_0^\infty h(\eta)n(t-\eta)\mathrm{d}\eta$$

so that
$$z(t) = \int_0^\infty g(\eta)u(t-\eta)\mathrm{d}\eta + \int_0^\infty h(\eta)n(t-\eta)\mathrm{d}\eta$$

and
$$z(t+\tau) = \int_0^\infty g(\eta)u(t+\tau-\eta)\mathrm{d}\eta + \int_0^\infty h(\eta)n(t+\tau-\eta)\mathrm{d}\eta.$$

Then
$$R_{uz}(\tau) = \underset{T\to\infty}{\mathrm{Limit}}\frac{1}{2T}\int_{-T}^{T}\left[u(t)\int_0^\infty g(\eta)u(t+\tau-\eta)\mathrm{d}\eta \right.$$
$$\left. + u(t)\int_0^\infty h(\eta)u(t+\tau-\eta)\mathrm{d}\eta \right]\mathrm{d}t$$
$$= \int_0^\infty g(\eta)\left[\underset{T\to\infty}{\mathrm{Limit}}\frac{1}{2T}\int_{-T}^{T} u(t)u(t+\tau-\eta)\mathrm{d}t \right]\mathrm{d}\eta$$
$$+ \int_0^\infty h(\eta)\left[\underset{T\to\infty}{\mathrm{Limit}}\frac{1}{2T}\int_{-T}^{T} u(t)n(t+\tau-\eta)\mathrm{d}t \right]\mathrm{d}\eta.$$

From which
$$R_{uz}(\tau) = \int_0^\infty g(\eta)R_{uu}(\tau-\eta)\mathrm{d}\eta + \int_0^\infty h(\eta)R_{un}(\tau-\eta)\mathrm{d}\eta.$$

The first term on the r.h.s. is the same as equation (5.22) and the second term is a convolution integral between $h(\eta)$ and the cross-correlation function of u with n. However, if u and n are uncorrelated $R_{un}(\ \ )$ is zero and

$$R_{uz}(\tau) = \int_0^\infty g(\eta)R_{uu}(\tau-\eta)\mathrm{d}\eta$$

from which we conclude that uncorrelated additive noise does not interfere with the measurement.

If we arrange that u is 'white' then this equation reduces to

$$R_{uz}(\tau) = Ng(\tau) \qquad \text{as before.}$$

Even with this property of noise immunity, the method as described, is idealised: we assume that we have infinite time to perform the experiment and that there are no quantisation errors. If we are using 'white' excitation we are assuming it is ideal. In reality none of these is true. However, we can approach the ideal as closely as is necessary for the work in hand; the time can be as long as necessary; we can reduce quantisation error by the choice of hardware and we can design the 'white' signal appropriately.

The application of these ideas is illustrated in two simple case studies, the first is rather artificial but is designed to bring out the essential features without too much difficulty and the second is a real application.

9.2 A Second Order System—Electronic Simulation

Using a simple analogue simulator a system was patched up having a transfer function of the form

$$G(s) = \frac{10^5}{s^2 + 251s + (628)^2}.$$

We easily deduce that $\omega_n = 628$ rad s^{-1}, (100 Hz) and $\zeta = 0.2$. The configuration is as shown in Fig. 9.2.

The input, u, with r.m.s. value 3.16 V, was derived from a generator and was a signal with a Gaussian PDD, the bandwidth was 500 Hz and thus approximates to a 'white' signal since the natural frequency of the system is $\frac{1}{5}$ of the bandwidth. Both u and y were applied to a correlator from which records were taken, the correlation information was also applied to a spectrum display unit which found the Fourier transform and thus displayed frequency information. Fig. 9.3 shows the arrangement. Both the correlator and spectrum display unit are digital instruments and their outputs are displayed by a cathode Ray oscilloscope (CRO) from which photographs can be taken. Fig. 9.4 shows the positive τ range of $R_{uu}(\tau)$ and is seen to be a narrow pulse which can be regarded as an approximation to an impulse. Its FT, that is, its spectral density $S_{uu}(\omega)$, is shown in Fig. 9.5 from which it is seen to be flat up to a break frequency of 500 Hz and thus approximates to a 'white' signal. Whilst not essential to this study it was easy to observe $R_{yy}(\tau)$ and $S_{yy}(\omega)$ and they are shown in Figs. 9.6 and 9.7. Regarding the system as a band pass filter with a 'white' input, we would expect the

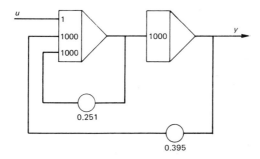

Fig. 9.2

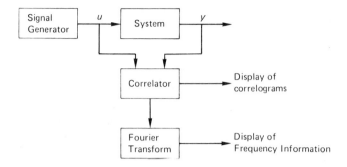

Fig. 9.3

Fig. 9.4

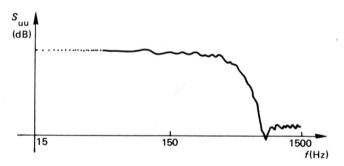

Fig. 9.5

Fig. 9.6

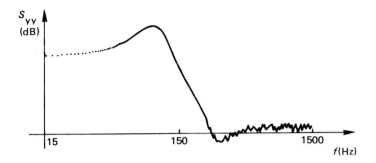

Fig. 9.7

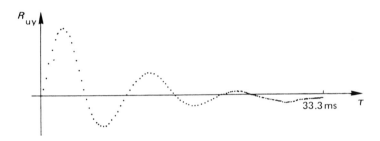

Fig. 9.8

output to display band limited characteristics, albeit not very narrow, and this is confirmed by $S_{yy}(\omega)$. Recall that the inverse FT of a band pass characteristic is an exponential cosine and this is found in $R_{yy}(\tau)$.

$R_{uy}(\tau)$ is shown in Fig. 9.8 and clearly represents an underdamped system and appears to be of the form

$$R_{uy}(\tau) = K e^{-t/T} \sin \omega_0 t = K e - \zeta \omega_n t \sin(\omega_n \sqrt{1 - \zeta^2})t \tag{9.1}$$

on assuming second order form. The identification problem is to find ζ, ω_n and the constant K where K is a gain term, ζ is the damping factor, and ω_n is the undamped natural frequency.

The time interval between each point is 333 μs and over two periods of the damped sinusoid we count 61.5 intervals which gives a periodic time,

$$T_0 = \frac{61.5 \times 333 \times 10^{-6}}{2}$$

$$= 0.010\,24 \text{ sec.}$$

This gives a natural frequency of

$$f_0 = \frac{1}{T_0} = 97.66 \text{ Hz}$$

$$\approx 98 \text{ Hz}$$

$$\omega_0 = 613.6 \text{ rad s}^{-1}$$

$$\approx 614 \text{ Hz.}$$

To find the time constant of the decrement, T, we use the fact that successive peaks of the damped sinusoid are related by

$$R_{i+1} = R_i e^{-\Delta t/T} \qquad \Delta t = \frac{\text{Time for one Period}}{2}$$

in which R_i is the height of a peak and Δt is the time interval between them. Re-arranging we get

$$T = \frac{\Delta t}{\log_e R_i/R_{i+1}}$$

and direct scaling of the first three peaks yields

$$T \simeq 9.32 \text{ ms.}$$

From equation (9.1)

$$\omega_0 = \omega_n \sqrt{1 - \zeta^2}$$

and

$$\frac{1}{T} = \zeta \omega_n$$

gives

$$\zeta = \sqrt{\frac{1}{T^2 \omega_0^2 + 1}}.$$

Substituting the above values gives

$$\zeta \simeq 0.172.$$

The theoretical values, of course, are

$$\omega_n = 628 \text{ r s}^{-1}$$
$$\omega_0 = 615 \text{ r s}^{-1}$$
$$f_0 = 97.97 \text{ Hz}$$
$$\zeta = 0.2.$$

The sources of errors are

(1) inaccuracies of the original simulation,
(2) inaccuracies in the correlator and
(3) inaccuracies in scaling from the diagram.

The problem of errors could be investigated in detail but this is not within the scope of this example.

We now have to find K and the simplest way is as follows.*

The first maximum occurs at $t = \dfrac{T_0}{4} = \dfrac{0.010\,24}{4} = 0.002\,56$ s

and has a magnitude $= K e^{-T_0/4T} = K e^{-0.002\,56/0.009\,32} = K \times 0.76$.

Scaling the magnitude from the diagram we get $1.1\ V^2$, that is

$$1.1 = K \times 0.76$$

so that

$$K = 1.447.$$

---

* A better way would be to measure K using a steady state method.

The parameter K is N multiplied by the gain term of $g(\tau)$ because $R_{uy}(\tau) = Ng(\tau)$. Let this gain term be K' then

$$K = NK'.$$

If we determine N we can easily find K'.

In section 5.13 we found that N is dimensionally (signal quantity)$^2 \times$ Time and is the strength (or area) of the impulse $R_{uu}(\ )$. Now Fig. 9.4 shows $R_{uu}(\tau)$ for positive time and since we have an even function we redraw it as Fig. 9.9 for clarification. If we approximate this to a triangle of height $10\ V^2$ and base $6 \times 333\ \mu s$ we get $N = 0.01\ V^2$ s.

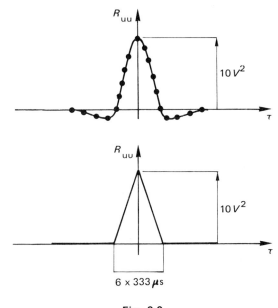

Fig. 9.9

Thus

$$1.447 = 0.01\ K'$$
$$K' = 144.7$$
$$\approx 145.$$

Based on our assumption that $R_{uy}(\ )$ is an exponentially damped sinusoid and that $R_{uy}(\ ) = Ng(\ )$ we conclude that $g(\tau)$ *is*

$$g(\tau) \simeq 145\mathrm{e}^{-t/0.009\,32}\sin 614t.$$

Since

$$G(s) = \frac{10^5}{s^2 + 251s + (628)^2}$$

$$g(t) = 162.5\mathrm{e}^{-t/0.007\,97}\sin 615.3t.$$

The accuracy of the measurement is not high by normal standards but since the spirit of the study has been to illustrate the principles involved no attempt has been made to invoke sophisticated observational and statistical methods which would have improved the accuracy.

We will now review the frequency response information. Recall from Chapter 4 that $\mathscr{F}[R_{uu}(\tau)]=S_{uu}(\omega)$ and had our input been truly 'white' $R_{uu}(\tau)$ would have been a true inpulse and $S_{uu}(\omega)$ would have been perfectly 'flat'. However, this was not so and $R_{uu}(\tau)$ approximated to a triangle of area 0.01 V^2s. Correspondingly $S_{uu}(\omega)$ was band limited and since

$$\overline{u^2} = \frac{1}{2\pi}\int_{-\infty}^{\infty} S_{uu}(\omega)d\omega$$

that is, the area under the $S_{uu}(\omega)$ curve is the mean square value (Fig 9.10), we can find the height of the $S_{uu}(\omega)$ curve. On linear scales for both S_{uu} and ω the curve cuts the ordinate at the bandwidth at half height and the curve is symmetrical so that the rectangle (dotted) is equal in area to that under the curve. The area is 10 V^2, its base is $2 \times 500 = 1000$ Hz so its height is $\dfrac{10}{1000} = 0.01$ V^2 seconds.

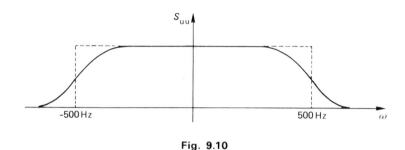

Fig. 9.10

It is no coincidence that we use the same symbol, N, for the strength of the R_{uu} impulse and for the height of the S_{uu} curve; they are, of course, one and the same thing.

In practice, all the signal power appears in positive frequency, thus $N=0.02$ V^2 seconds. Fig. 9.5 shows S_{uu} on an axis of $\log \omega$ with the ordinates measured in dB and the flat form over the working range of frequencies is apparent.

Fig. 9.7 is S_{yy} ($\log \omega - $dB) and scaling from the record gives a peak height above low frequency level of approximately 7.5 dB which tallies well with standard curves for this damping factor. The high frequency slope is approximately 40 dB per decade. (The ragged flat region at the right hand end of the record is inaccurate and subject to instrumentation limitations.) The peak is at approximately 100 Hz.

Fig. 9.11 shows the Bode diagram pair (gain and phase) for the transfer function and are, of course, characteristic of that for a second order system and are obtained by taking the FT of $R_{uy}(\tau)$ that is $S_{uy}(\omega)=\mathscr{F}[R_{uy}(\tau)]$. $S_{uy}(\omega)$ is, of course, complex and thus has real and imaginary components which give rise to the gain and phase diagrams. The gain curve is $10\log_{10}|S_{uy}(\omega)|$ not $20\log_{10}|S_{uy}(\omega)|$ and due allowance is made in interpretation. The peak is at approximately 100 Hz.

Fig. 9.12 shows $S_{uy}(\omega)$ on polar axes and is again a shape characteristic of a second order system.

All this spectral data was obtained with the instrumentation configured to use a Bartlett window on the correlogram data. It is interesting to note the effect when a rectangular window is used and Figs. 9.13 and 9.14 show $R_{uy}(\tau)$ and $S_{uy}(\omega)$ (polar) in this configuration. The effect of the side lobes is apparent and severe distortion is obvious.

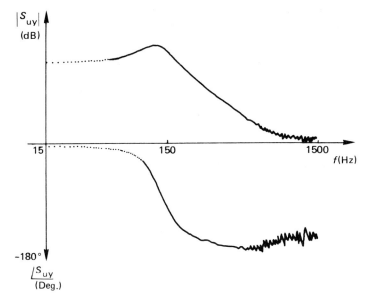

Fig. 9.11

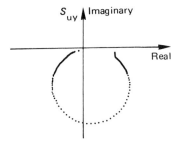

Fig. 9.12

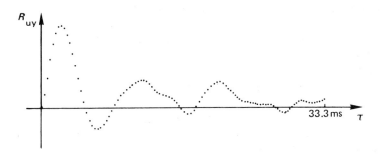

Fig. 9.13

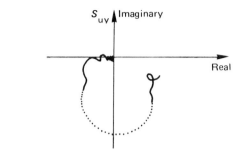

Fig. 9.14

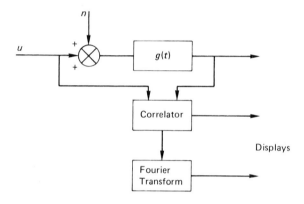

Fig. 9.15

The above tests (using a Bartlett window) were repeated with the addition of noise interference as shown in Fig. 9.15. The noise was narrow band with centre at 500 Hz and bandwidth 300 Hz. The upper trace in Fig. 9.16 shows u and the lower shows $u+n$. Figs. 9.17 and 9.18 show R_{yy} and S_{yy} respectively, and as expected the interfering noise shows its effect, especially in S_{yy} where a peak at 500 Hz is evident. Figs. 9.19, 9.20 and 9.21 show, respectively R_{uy}, S_{uy} and S_{uy} (polar) and are almost indistinguishable from those obtained without noise which demonstrates the noise immunity of the method.

To emphasise the importance of the proper choice of input signal, u, for such identification tests, we show the effect of using a signal with too narrow a bandwidth. The original test, that is, without noise interference and using a Bartlett window, was repeated but u (again gaussian in distribution) had a bandwidth of 50 Hz. Fig. 9.22 is its R_{uu} and Fig. 9.23 is its S_{uu} from which it is easily seen, when compared with Figs. 9.4 and 9.5 that the bandwidth is narrower and the auto-correlation function is wider. Obviously the 'tail' in R_{uu} is comparable with that of g the impulse response and similarly, the cut-off frequency of S_{uu} is lower than the undamped natural frequency of g. Fig. 9.24 shows R_{uy} and Fig. 9.25 shows S_{uy} (polar) and there is no resemblance whatsoever with those of Figs. 9.8 and 9.12. Clearly, this test yields results which gives conclusions which are grossly incorrect.

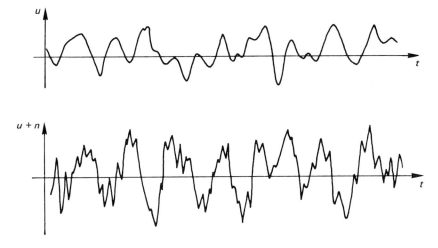

Fig. 9.16

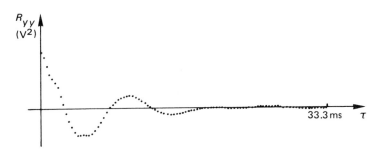

Fig. 9.17

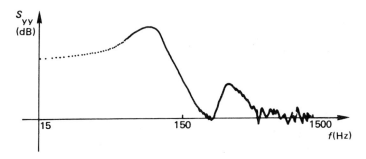

Fig. 9.18

9.3 The Use of Pseudo Random Binary Signals (PRBS)

The main criterion for the input signal for systems under identification is that its auto-correlation function should be a narrow impulse which implies automatically that its bandwidth should be wide. A narrow impulse (or wide bandwidth) is relative and depends

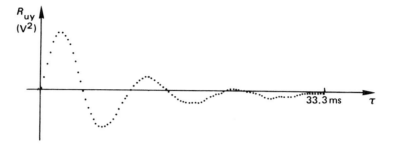

Fig. 9.19

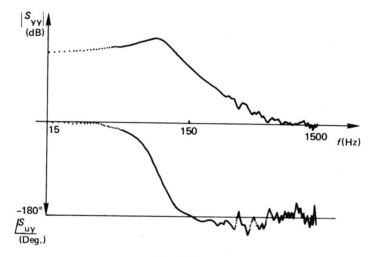

Fig. 9.20

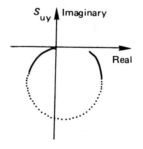

Fig. 9.21

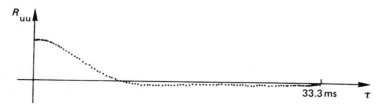

Fig. 9.22

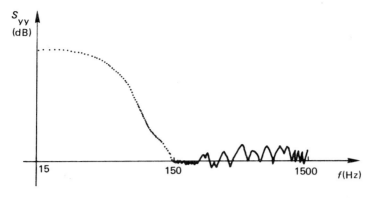

Fig. 9.23

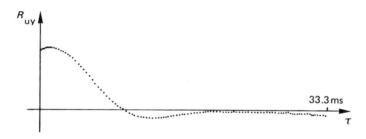

Fig. 9.24

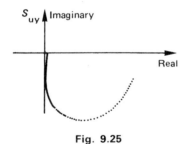

Fig. 9.25

on the time constants and natural frequencies of the system under identification. The probability distribution of the input does not enter the argument, thus we could look for other signals, that is, non-gaussian ones, that have the proper form of auto-correlation.

A class of signals which are candidates for use in system identification are binary signals and these are ones with only two admissible levels (like a square wave). A characteristic which makes them candidates is that of ease of generation along with the fact that they are often easy to apply to a system, this makes them very attractive. For instance, it is easy to apply a square wave to a high power electrical machine by simply opening and closing a pair of contacts between the supply and machine (provided adequate precautions are taken to keep the resulting current changes to reasonable proportions). However, the familiar equal mark-to-space ratio square wave is unsuitable because its auto-correlation is not an impulse. (See section 3.8). If we take a square wave in which the transitions from one level to the other are random we find[15] that its R_{uu} is exponential of the form shown in Fig. 9.26 and if the time constant of R_{uu} is chosen to be very small in comparison with the shortest one in the system under identification then we have a usable signal.

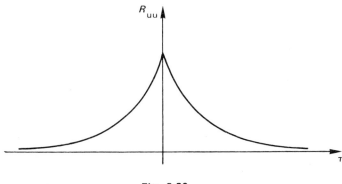

Fig. 9.26

A class of binary signals is that named pseudo random binary[15]. They are very easy to generate using shift registers but, strictly speaking, they are deterministic because the transitions from one level to the other only occur at clocked times and are periodic. A generator designed to produce these signals would have a clock generator built in whose periodicity could be varied from, say, 0.01 Hz to several MHz and a means whereby the sequence length could be varied. Fig. 9.27 shows a typical PRBS from which it will be seen that the pattern of transitions is repeated at intervals of seven clock periods, making 7 the sequence length. It is not within the scope of this book to study the details of the generation and properties of these signals but it is found that some admissible lengths are 7, 15, 31, 63, 127, 255, 511 clock periods. These numbers are found from $N = 2^n - 1$ where n is 3, 4, 5, 6, (i.e. integers larger than 2).

It can be shown that the auto-correlation function of this class of signals is as shown in Fig. 9.28. Note that it is periodic as one would expect since the signal is periodic. Now if, for a given identification problem, the clock period is set to about $\frac{1}{5}$ of the shortest expected time constant of the system under test and the sequence length [in seconds, i.e. clock period $\times (2^n - 1)$] is arranged to be about 5 times the longest time constant then R_{uu} seems to the system to be a single impulse at the time origin.

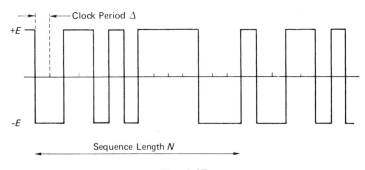

Fig. 9.27

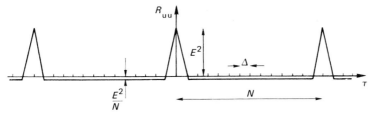

Fig. 9.28

Figs. 9.29 to 9.33 show the results of a test on the simulated second order system of Fig. 9.2 using the arrangement of Fig. 9.3 with a PRBS as the input, u. The clock period was 100 μs and the sequence length was 511, (i.e. $511 \times 100 \, \mu s = 51.1$ ms).

These results are indistinguishable from those obtained using a gaussian signal as can be seen when Figs 9.29 to 9.33 are compared with Figs. 9.8, 9.11 and 9.12. This demonstrates clearly the validity of PRBS as a perturbation signal for system identification.

As expected if the PRBS is not designed properly then spurious results are obtained. Figs. 9.34 to 9.36 show R_{uy} and S_{uy} resulting from a repeat of the previous test using a sequence length of 127, (i.e. 12.7 ms). R_{uy} shows not only the expected damped sine wave but the beginning of a second one resulting from the second auto-correlation impulse. Its effect on S_{uy} and hence on the estimate of $g(t)$ [or $G(s)$] is disastrous.

If the clock period had been too long then the effect on the results would have been similar to that obtained when the test using a gaussian signal with too narrow a bandwidth was employed. See Figures 9.22 et seq.

Fig. 9.29

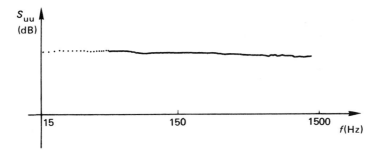

Fig. 9.30

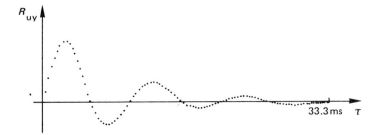

Fig. 9.31

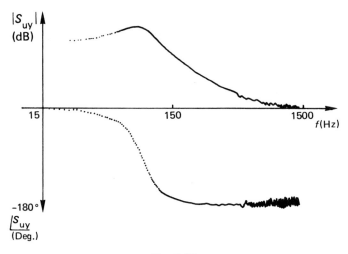

Fig. 9.32

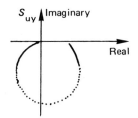

Fig. 9.33

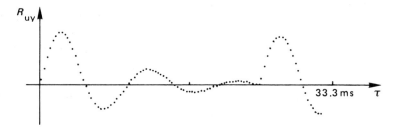

Fig. 9.34

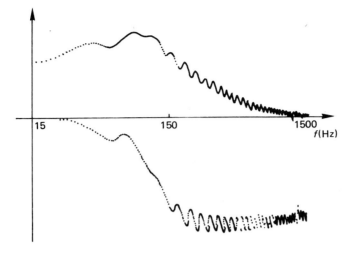

Fig. 9.35

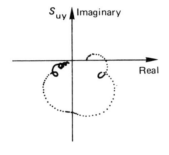

Fig. 9.36

9.4 Identification of a D.C. Motor[16]

We now look at the results obtained from an identification test on a 2 kW 110 V d.c. motor. The machine was supplied with a separate field excitation which was maintained constant throughout. The armature was supplied from a constant voltage source of 110 V via a power transistor whose base was supplied with a gaussian signal. This arrangement allowed the armature to be perturbed in a gaussian manner about its normal operating value which in turn caused the speed to be perturbed about its normal operating value of 1320 r.p.m. which was observed by a speed to voltage transducer.

The gaussian perturbation at the transistor collector was regarded as input, u, and the speed perturbation was output, y. The normal operating values of voltage and speed were backed off so as not to enter the correlator and give a bias on the correllograms. Figs. 9.37 to 9.40 show the results of the identification test for which the bandwidth of the gaussian signal was 50 Hz.

Using the same approach as in the simulations, that is, scaling from the correllogram the impulse response of the system was found to be

$$g(t) = 413e^{-t/1.1}.$$

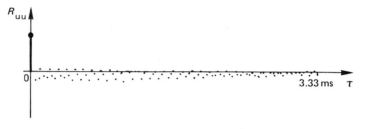

Fig. 9.37

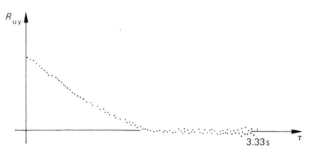

Fig. 9.38

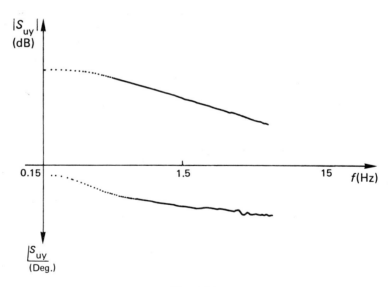

Fig. 9.39

However, this includes the transistor into the system; in particular its output resistance is effectively in series with the armature resistance which, of course, affects both the gain (413) and the time constant (1.1s). To arrive at the relevant motor parameters clearly requires correction using the output resistance of the transistor and electrical circuit analysis which is not within the scope of this book.

An interesting, but incidental, effect can be noticed on examining R_{uu} where it will be seen that the 'tail' of the correllogram is ragged. This is an aliased representation of a periodicity on the normal operating voltage of the armature which arises due to imperfect smoothing of the rectified power supply. Its frequency and amplitude hardly affect the speed at all.

In the case studies cited above the form of the correllograms, and hence of the relevant impulse responses, was well defined. At first we expected it to be in the form of an underdamped second order system because of our knowledge of the simulation and in the second we were not surprised to find a first order system since a d.c. motor behaves as such. In more complicated cases the form is not nearly so evident. In subsequent work on the d.c. motor more detailed investigations were followed where an attempt was made to find the armature inductance time constant; thus the model was now regarded as second-order but one in which the values of the two time constants were quite different and little knowledge was available regarding their relationship. Having obtained a cross-correllogram its parameters were estimated by what is known as a model reference method. This uses a computer to compare the observed correllogram with an assumed one, the parameters of the latter are adjusted automatically until a best fit is obtained. The details of this method will not be discussed here but they will be mentioned to indicate extensions of the basic method[15].

9.5 Signal Analysis Applied to Vibrating Systems— A Vibrating Beam

One area of study in which signal processing techniques are used extensively is that associated with vibrations[17]. Most people are familiar with some aspect of vibrations, e.g. a car suspension system, a vibrating car body panel or the oscillations of a washing machine or spin-drier as it increases in speed to its desired operating speed.

In a large number of cases the vibrations are unwanted and engineers spend a considerable amount of time trying to reduce or minimise their effects. This is particularly true in the case of rotating machinery when the cause of a vibration may be some imbalance in the system which in turn may cause bearing wear and even failure. An investigation of the vibrations produced by such a system using signal processing techniques, such as spectral analysis, can indicate the source of the problem.

However, in this particular investigation we are going to use random signals to investigate the behaviour of a simple vibrating structure. This method of testing structures is used in a variety of industries. Once the characteristics of the structure are known then measures can be introduced to reduce the levels of vibration during normal operation. An example of this is in car body design. Modern vehicle design can lead to noise problems inside the vehicle due to the vibration of the body. The excitation of the structure is due to a combination of aerodynamic forces and forces transmitted to the body via suspension system. In order to reduce the vibration levels, which give rise to the noise, the vehicle bodies are subjected to extensive vibration testing during the design of the vehicle. The results of this type of study help to identify the parts of the body which cause the major problems and hence those which need some redesign or added damping to reduce the overall vibration.

Our structure is not as complex as a car body, nevertheless, the principles involved are the same and similar problems arise in the interpretation of the data. The structure is a simple beam fixed at either end to solid supports, see Fig. 9.41.

The beam is excited by means of an electro-mechanical shaker driven from a noise generator via a power amplifier. The mechanical link between the shaker and the beam has

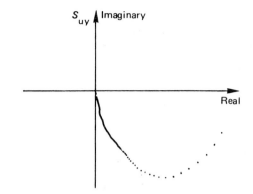

Fig. 9.40

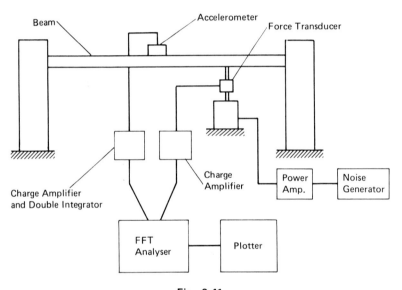

Fig. 9.41

a force transducer in-line. The resultant motion of the beam is sensed by an accelerometer. Both the force transducer and the accelerometer are piezo-electric transducers, thus the signals must be passed through a charge amplifier to produce signals which match the input to the analyser. In the case of the accelerometer signal the charge amplifier also provides a double integration facility to enable the displacement of the beam to be used in the analysis.

For the purpose of this test the beam is assumed to be a linear system, therefore we can apply the theory developed in Chapter 6 in order to determine the frequency response characteristics of the beam, i.e. the transfer function. Consider the block diagram representation of the system shown in Fig. 9.42.

From this we can see that it is possible to get more than one transfer function for the system. The most obvious transfer function and, indeed the desired transfer function, is that obtained from a combination of the force and displacement signals. This pre-supposes that the conditioning instrumentation, the charge amplifiers in this case, have been selected to

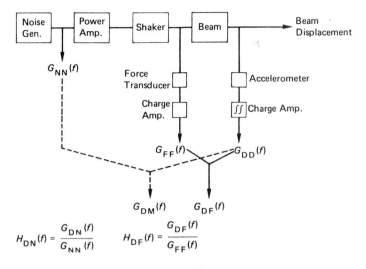

$$H_{DN}(f) = \frac{G_{DN}(f)}{G_{NN}(f)} \qquad H_{DF}(f) = \frac{G_{DF}(f)}{G_{FF}(f)}$$

Fig. 9.42

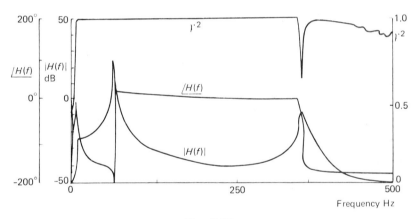

Fig. 9.43

minimise their influence on the measured response. This requirement is satisfied when the frequency characteristics of the instrument are such that the signals are not modified in amplitude over the frequency range of interest, other than by a constant scale factor, and the phase of the signals are not changed in relation to each other.

The first set of results shows the frequency characteristics of the beam in the frequency range 0–500 Hz, Fig. 9.43. From the results we can see that there are two natural frequencies, i.e. two modes of vibration, within this frequency range. Associated with each resonance peak is the 180° change in phase. An equally important parameter to be investigated, from the point of view of the reliability of the results, is the coherence function. This gives a measure of the dependence of the system output on the input. For a linear system this should result in a value of unity over the entire frequency range, i.e. the output is totally dependent on the input. However, the results show that the coherence drops below unity at low frequencies, at the second resonance and again above the second resonance

(> 370 Hz). The low coherence at either end of the frequency range is due mainly to the inadequacy of the excitation of the system. The low coherence value at the second resonance is due to the reduced sensitivity of the analysis. The usefulness of the coherence function will now to be demonstrated.

To overcome some of the problems, it is often convenient to concentrate on one mode of vibration at a time. This can be done using 'zoom' analysis which enables a specified band of frequencies to be analysed rather than to half the sampling frequency as dictated by conventional Fourier analysis (section 7.8). A more detailed examination of the first mode of vibration yields the results shown in Fig. 9.44 where the frequency range is 28.6 Hz–130.2 Hz.

The amplitude characteristics now yield a single resonance peak at a frequency of 68.8 Hz, but the top of the peak is slightly flattened indicating that the true resonance peak should be of a larger amplitude and at a different frequency. The coherence function also falls rapidly at resonance, to 0.7, indicating that the analysis is in error, most probably due to poor resolution. This is confirmed when the frequency range is shifted slightly (29.6 Hz to

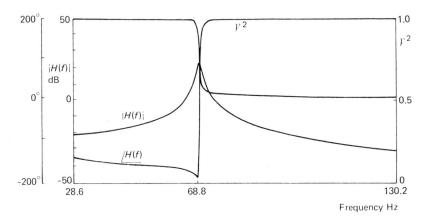

Fig. **9.44**

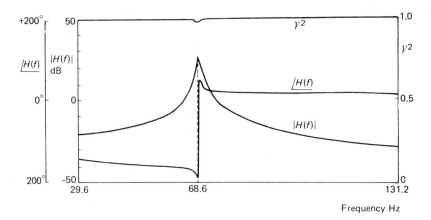

Fig. **9.45**

131.2 Hz), Fig. 9.45. We now see that a well defined resonance peak is produced with only a slight dip in the coherence function at resonance.

In both sets of tests the data was windowed and averaged (32 ensemble averages used) and the frequency resolution was 0.8 Hz. The improvement in the second set of results is due to the resonance peak now being more closely aligned with a computed spectral line which reduces the resolution error and improves the coherence function.

A commercial FFT analyser was used so that the sample rate and anti-aliasing filters are automatically selected to suit the frequency range of the analysis. Care should be taken to ensure that the inputs to the analyser are set to the most sensitive settings without overloading the input amplifiers. This helps to reduce errors in the analysis.

The amplitude and phase spectra of the first mode of vibration, when the analysis is performed on the noise and beam displacement signals, are shown in Fig. 9.46. The amplitude spectrum now exhibits the characteristics of a more heavily damped system than the displacement—force spectrum. The transfer function is correct for the system analysed, but of course we have investigated a different system and most importantly used a different input signal (the noise generator output rather than the force) as the basis of the analysis.

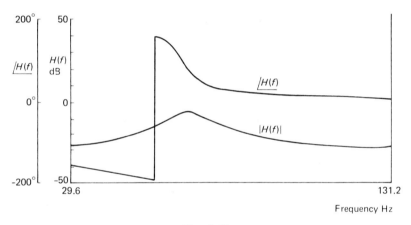

Fig. 9.46

This rather obvious difference between the two sets of data has been used to draw the readers attention to the necessity to have an understanding of the system being investigated, otherwise it is possible to obtain very accurate and precise results which have little physical meaning.

9.6 Signal Analysis Applied to Vibrating Systems— A Large Industrial Mixing Vessel[13]

This investigation illustrates the way in which the techniques of signal processing and system identification, which have been discussed, can be applied to gain information about the behaviour of a system which would not otherwise be readily available.

The problem which was investigated was concerned with determining the condition of the agitator in a large industrial mixing vessel, Fig. 9.47 without making intrusive measurements. The parameters of importance, which influence the operation of the agitator, are the force and vibration levels experienced by the impeller and shaft. An inspection of Fig. 9.47

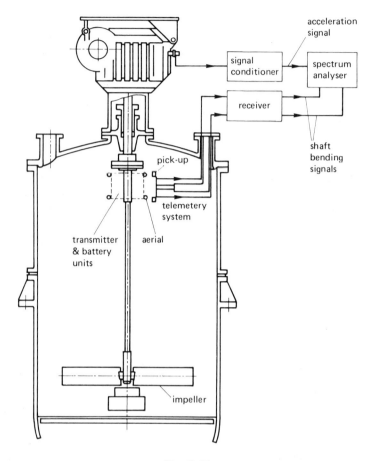

Fig. 9.47

will indicate the reason why it is important to have information about these parameters since the shaft is only supported by the top bearings, therefore any overloading of the agitator will directly affect the life of the bearings and hence increase the frequency of maintenance on the system. This in turn increases the operating costs of the plant.

The requirement is to be able to predict the forces experienced by the impeller from non-intrusive vibration measurements made at the bearing housing. The technique was developed in two stages:

(1)   determine the forces acting on the impeller, and
(2)   measure the vibrations at the bearing housing and relate those to the measured forces.

Clearly stage (1) is only required to verify the technique but does itself require an interesting solution to a measurement problem since the measurements are made whilst the mixing vessel is in operation and the forces cannot be measured directly.

The first part of the work required specialised instrumentation to extract the signals from the rotating machinery. A telemetry system was used which converts the voltage, obtained from strain gauges monitoring the impeller shaft strain, into a frequency modulated signal. This signal is then transmitted to a receiver and a demodulator is used to convert the signal back to a voltage which can then be related to the strain in the impeller shaft and hence the stress. However, knowing the stress levels in the impeller shaft does not give an accurate indication of the loading applied at the impeller. This is because the stress due to a particular loading condition is the response of a system to a force input. Thus, the shaft and impeller are a system and can be modelled as shown in Fig. 9.48. Similarly the instrumentation-telemetery system modifies the observed strain signal so that this process can also be represented by a similar arrangement, Fig. 9.49.

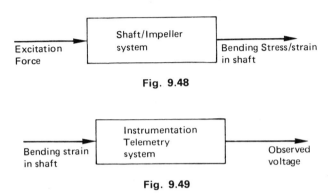

Fig. 9.48

Fig. 9.49

Therefore if the characteristics of the two modifying systems can be determined, then the excitation force can be derived from the observed voltage. The required characteristic of each system is the transfer function expressed in terms of frequency. In the case of the instrumentation system this can be measured by applying known excitation signals and measuring the response; therefore the transfer function can be obtained as shown in the previous section.

The transfer function of the shaft was obtained by analysis. A mathematical model was derived for the shaft using the transfer matrix technique[12] and the model then expressed in terms of frequency. Thus, both intermediate systems can be described in terms of frequency response characteristics, and hence it is possible to determine the excitation forces acting on the agitator from a knowledge of the observed voltage. The procedure for doing this is shown in Fig. 9.50. Thus:

$$E(f) = H_s(f) . H_I(f) . V(f)$$

where $E(f)$ is the force spectrum
   $H_s(f)$ is the shaft/impeller transfer function
   $H_I(f)$ is the instrumentation transfer function
   $V(f)$ is the 'observed voltage' spectrum.

The results for a typical stress spectrum are shown in Fig. 9.51 together with the transfer functions for the intermediate systems.

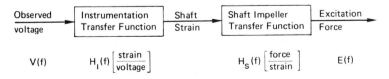

Fig. 9.50

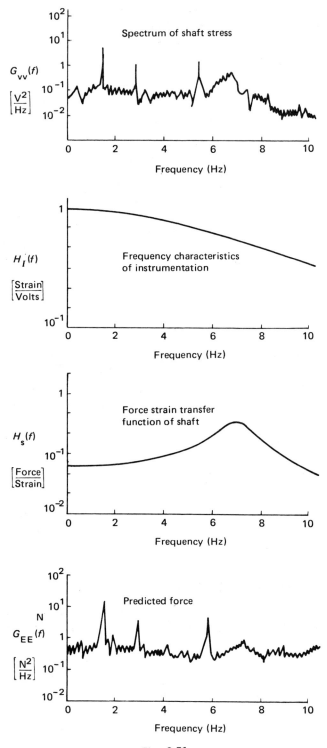

Fig. 9.51

Although many simplifying assumptions have been made in the derivation of the excitation spectrum, the model of the agitator is based, for example, on linear theory, nevertheless the results are representative of the excitation forces acting on the agitator and the relative magnitudes of the force at the dominant frequencies in the spectrum give a clear indication of the speeds to be avoided when selecting suitable operating conditions for the mixer.

The second stage is to relate the vibrations monitored at the bearing housing to the force on the impeller. Once again this requires two items of information:

(1) the relationship between the vibration and exciting forces, and
(2) the frequency characteristics of the instrumentation.

Item (1) was the subject of a research programme and resulted in the development of a modelling technique which adequately represented the vibration/force characteristics of impeller, shaft and associated supporting structure of this type of plant.

The instrumentation included an accelerometer, charge amplifier and a filter incorporating a double integration facility so that the acceleration signal could be converted to displacement. The frequency characteristics of this instrumentation were obtained in a similar manner to the previous test.

The signal processing required to determine the impeller force can now be represented by the block diagram shown in Fig. 9.52.

The results of a typical test are shown in Fig. 9.53(b) which can be compared with the direct estimate of the force spectrum, Fig. 9.53(a). The results compare very favourably, up to a frequency of 7 Hz, with both the amplitude and frequency of the predicted force levels being in close agreement for both methods of measurement. The more direct measurement of the force, i.e. by measuring the impeller shaft stress, is assumed to be an accurate prediction of the impeller forces, therefore we see that the non-intrusive measurement technique provides a satisfactory basis for determining the force levels experienced by the agitator.

The analysis, including all the spectral estimation, was done using a general purpose 16-bit microcomputer and specially developed software. A standard FFT algorithm was used and all the time data was windowed (Hanning window) before the transform was applied. A high sample rate was used throughout the tests so that aliasing was not a problem since that frequency range of interest was low (10 Hz maximum). A 16 channel 12-bit analogue to digital converter with a voltage range of ± 10 V was used to log the data and all the signals were scaled to ensure that the full input range of the ADC was used, thereby reducing the effects of quantisation error.

9.7 Signals and Noise in Fluid Dynamics

In real applications the signals to be investigated are often corrupted by noise. In Chapter 8 we saw how filtering techniques can be used to remove unwanted noise but in many

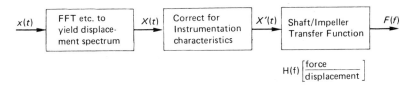

Fig. 9.52

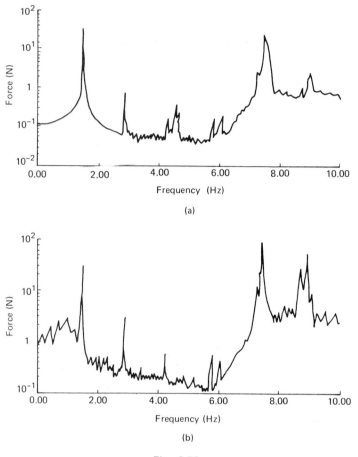

(a)

(b)

Fig. 9.53

instances the noise on the signal is an inherent part of the process which gives rise to the signal. A typical application where such an effect can be observed is in the study of 'vortex shedding' associated with flow around solid objects. The results which are described are from a laboratory experiment but the phenomenon has considerable practical importance since the shedding of the vortices gives rise to a periodic transverse force on the object which can lead to fatigue failure. The effects can be catastrophic and expensive when the resultant failure is a suspension bridge, boiler tube, chimney or electricity transmission cable.

The test involves passing air around a solid object and measuring the fluid velocity downstream from the object, see Fig. 9.54. The test object is a cylinder of rectangular cross-section which is placed in a wind tunnel and the resultant fluctuating fluid velocity is measured using a hot wire anemometer. This is an extremely sensitive transducer which operates on the principle of the air flow cooling a very fine heated wire and the cooling effect can be related to the velocity of the air flow. This type of transducer has extremely good frequency response characteristics and is capable of responding to velocity fluctuations > 1 kHz. The test was performed with an air flow of 13.9 m s^{-1} around the cylinder and the objective of the test is to determine the vortex shedding frequency for this particular configuration.

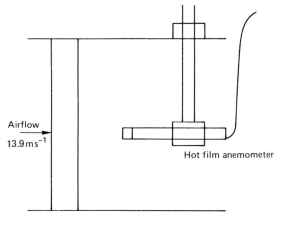

Fig. 9.54

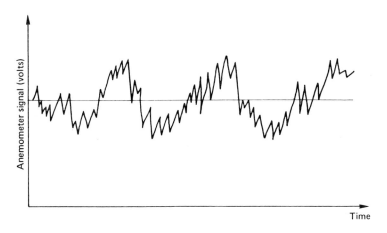

Fig. 9.55

Inspection of the time data, Fig. 9.55, indicates the presence of a periodic component in the signal but this is masked to a large extent by noise. In this case, the noise is due to the turbulence of the flow which is associated with vortex shedding. The time signal is so corrupted by the noise that it is clearly not possible to determine the period of the dominant component from this signal. The question which then arises is, how can the information be extracted from this signal?

We are only interested in determining the frequency of the periodic component, therefore two methods of analysis are equally applicable; correlation and spectral analysis. The signal could also be filtered to extract the information, using a band pass filter, but as we shall see the other two methods provide a quick and reliable method of determining the required frequency.

The auto-correlation function of the signal is shown in Fig. 9.56. This clearly indicates the nature of the periodic component and the periodic time can be measured directly (8.8 ms) from which the shedding frequency is 113.6 Hz. The noise on the signal is characterised by

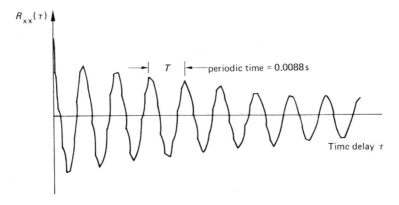

Fig. 9.56

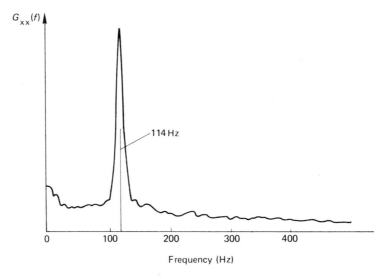

Fig. 9.57

the initial 'spike' in the auto-correlation function. The decay in the amplitude of the periodic component is indicative of some other transient fluid flow phenomenon.

The frequency spectrum of the signal, Fig. 9.57 again shows that the periodic component dominates the result with the noise exhibiting the spectral characteristics of turbulence, i.e. a reduction in amplitude with increased frequency. The resolution of the frequency analysis (2 Hz) yields a shedding frequency of 114 Hz which is in good agreement with that obtained from the correlation method.

This relatively simple test indicates the 'power' of signal processing techniques to extract information from what appears to be extremely poorly measured data.

References

1. Bendat, J. S. and Piersol, A. G., *Random Data: Analysis and Measurement Procedures*, John Wiley, New York, 1971
2. Ogata, K., *Modern Control Engineering*, Prentice-Hall, 1970
3. Cooley, J. W. and Tukey, J. W., *An Algorithm for the Machine Calculation of Complex Fourier Series*, Mathematics of Computation, 1965, **19**, 297–330.
4. Harris, F. J., On the Use of *Windows for Harmonic Analysis with the Discrete Fourier Transform*, *Proc. IEEE*, Jan 1978, **66**, No. 1, 51–83.
5. Durrani, T. S. and Nightingale, J. M., *Data Windows for Spectral Analysis, Proc. IEE*, 1972, **119**, No. 3, 343–52.
6. Newland, D. E., *An Introduction to Random Vibration and Spectral Analysis*, Longman, London, 1975.
7. Murdoch, J. and Barnes, J. A., *Statistical Tables for Science, Engineering, Management and Business Studies*, Macmillan, 1974
8. Blackman, R. B. and Tukey, J. W., *The Measurement of Power Spectra*, Dover, New York, 1959
9. Bozic, S. M., *Digital and Kalman Filtering*, Edward Arnold, London, 1971
10. Stanley, W. D., *Digital Signal Processing*, Prentice-Hall, 1975
11. Oppenheim, A. V. and Schaffer, R. W., *Digital Signal Processing*, Prentice-Hall
12. Thompson, W. T., *Vibration Theory and Applications*, Allen and Unwin, 1973
13. Kwok, C. C., *Vibration Monitoring of Mechanical Agitators*, PhD Thesis, University of Manchester, 1982
14. Schwarzenbach, J. and Gill, K. F., *System Modelling and Control*, 2nd Edition, Edward Arnold, 1984
15. Eykhoff, D., *System Identification*, Wiley, 1974.
16. Brook, D. and Morton, D., *Modern Techniques in Electrical Machine Parameter Identification*, IFAC, York, 1985.
17. Brook, D. and Wynne, R. J., *A Comparison of Time Domain and Frequency Domain Methods of Measuring Damping and Natural Frequencies of Structures*, IFAC, Darmstadt, 1979.
18. Doebelin, E. D., *Measurement Systems*, McGraw Hill, 1982.

Bibliography

Franklin, G. F. and Powell, J. D. *Digital control systems*, Addison-Wesley, 1980.

Lathi, B. P., *Random Signals and Communication Theory*, ITC, 1968

Lee, Y. W., *Statistical Theory of Communication*, Wiley, 1967

McGillern, C. D. and Cooper, G. R., *Continuous and Discrete Signal and System Analysis*, 2nd Edition, Holt-Rinehart and Winston, 1984

Thomas, J. B., *Statistical Communication Theory*, Wiley, 1969

Papoulis, A., *Circuits and Systems—A Modern Approach*, Holt Saunders, 1980

Papoulis, A., *Signal Analysis*, McGraw-Hill, 1977

Appendix A

1 The Laplace Transform

The Laplace transform has been defined in the text (Chapter 4) and used in the modelling aspects of Chapters 5, 8 and 9. The purpose of this short section on Laplace transforms is to summarise the essential features of the transform as revision for those who are returning to this topic after some time and to serve as a reference for those wishing to compare the Fourier and Laplace transforms. Those unfamiliar with Laplace transforms are recommended to refer to the many texts dealing with the subject. The less mathematically inclined should consult one of the many textbooks on Control Engineering which also cover the Laplace Transform theory to the required level[2, 14]

Definition

Given a function of time $f(t)$ the Laplace transform of this function is defined by the integral equation

$$F(s) = \int_0^\infty f(t)e^{-st}\,dt. \tag{A.1}$$

This defines a function of s provided the values of s cause the function to converge. The range of integration is for $t \geqslant 0$ and $f(t)$ is normally assumed to be zero for $t < 0$; s is a complex variable.

Note—The bilateral Laplace transform is defined as

$$F_b(s) = \int_{-\infty}^\infty f(t)e^{-st}\,dt \tag{A.2}$$

but this has only limited application as most real systems are defined for positive time only.

Example

(i)   Consider the function of time to be

$$\begin{aligned} f(t) &= 0 & t < 0 \\ f(t) &= 1 & t \geqslant 0 \end{aligned}$$

Fig. A.1

which is shown diagrammatically in Fig. A.1. This function is often called the unit step. The Laplace transform is given by application of equation (A.1)

$$F(s) = \int_0^\infty 1 . e^{-st} dt$$

$$= \left[-\frac{1}{s} e^{-st} \right]_0^\infty$$

$$= \left[0 - \left(-\frac{1}{s} \right) \right]$$

$$= \frac{1}{s}.$$

It was noted in Chapter 4 that useful transformations are reversible and the Laplace transform is a reversible process

$$f(t) \xleftarrow{\text{reversible}} F(s)$$

thus

$$1 \longleftrightarrow 1/s.$$

(ii) Consider next the function $f(t) = e^{at}$ where a can be either a positive or negative constant

$$F(s) = \int_0^\infty e^{at} . e^{-st} dt$$

$$= \int_0^\infty e^{-(s-a)t} dt$$

$$= \left[-\frac{e^{-(s-a)t}}{(s-a)} \right]_0^\infty$$

$$= \frac{1}{s-a}$$

Thus

$$e^{at} \longleftrightarrow \frac{1}{s-a}.$$

The condition for convergence of this particular transform is that $s > a$ (when $a > 0$). This is a necessary constraint since the function e^{at} is divergent and thus the transform must be bounded. When $a < 0$ the transform becomes

$$e^{-at} \longleftrightarrow \frac{1}{s+a}$$

and the condition for convergence is satisfied when the real part of s is positive. The general rule for covergence is that

$$f(t)e^{-st} \to 0 \qquad \text{as } t \to \infty.$$

Thus the limiting values of s for convergence can be determined.

By evaluating the transformation integral for commonly used functions $f(t)$ we can construct a table of transforms. Table A.2 is typical.

The Laplace Transform of Derivatives

The Laplace transform of a derivative $f'(t)$ is given by

$$F_d(s) = \int_0^\infty f'(t) e^{-st} \, dt \tag{A.3}$$

but we wish to express this transform in terms of $F(s)$ the transform of $f(t)$. Integrating equation (A.3) by parts yields

$$F_d(s) = [f(t) e^{-st}]_0^\infty + s \int_0^\infty f(t) e^{-st} \, dt$$

$$= -f(0) + sF(s).$$

Thus the transform of a derivative can be expressed in terms of the transform of the time function $f(t)$ and the value of the function at time $t = 0$. This term is called the initial condition.

The same principle can be applied to higher order derivatives. Thus

$$\int_0^\infty f''(t) e^{-st} \, dt = [f'(t) e^{-st}]_0^\infty + s \int_0^\infty f'(t) e^{-st} \, dt.$$

Substituting from the previous results yields

$$\int_0^\infty f''(t) e^{-st} \, dt = [f'(t) e^{-st}]_0^\infty + s[f(t) e^{-st}]_0^\infty + s^2 \int_0^\infty f(t) e^{-st} \, dt$$

$$= -f'(0) - sf(0) + s^2 F(s)$$

where $f'(0)$ is the value of the derivative of $f(t)$ at $t = 0$. Thus the general form of the Laplace transform of a derivative is given by

$$\int_0^\infty f^n(t) e^{-st} \, dt = -f^{n-1}(0) - sf^{n-2}(0) - s^2 f^{n-3}(0) - \ldots .$$

$$\ldots -s^{n-1} f(0) + s^n F(s). \tag{A.4}$$

Example

Laplace transforms are most often used to solve differential equations. Consider the differential equation given below which is to be Laplace transformed. Since $x(t)$ is not specified at this stage it is assumed that the transform $X(s)$ exists

$$3x'(t) + 2x(t) = 0$$

and
$$x(0) = 1.$$

Thus
$$\mathscr{L}[2x(t)] = 2X(s)$$

and
$$\mathscr{L}[3x'(t)] = 3[sX(s) - x(0)]$$

$$= 3sX(s) - 3:$$

Thus the transformed equation becomes:

$$3sX(s) + 2X(s) = 3.$$

We will see later how the time solution of this equation is obtained.

The Inverse Transform

So far we have assumed that the transformation procedure is reversible. The mechanism of inversion is, however, rather complex if approached from first principles since the resulting integral must be evaluated as a line integral in the complex s-plane $(\sigma, j\omega)$. To avoid this problem the inversion process is normally evaluated by reference to a table of standard transforms where $F(s)$ has been obtained for many of the time functions $f(t)$ which are most often required. By this means we can complete the solution of the differential equation of the previous example.

Example (continued)

$$3sX(s)+2X(s)=3$$

The transformed equation is now an algebraic equation in s and so it can be manipulated using the normal conventions of algebra. Thus

$$(3s+2)X(s)=3$$

or

$$X(s)=\frac{3}{3s+2}.$$

We have already shown that

$$F[e^{-at}]=\frac{1}{s+a}$$

Therefore, manipulating $X(s)$ gives

$$X(s)=\frac{1}{s+2/3}.$$

Thus the inverse transform $x(t)$ and hence the solution to the differential equation is given by

$$x(t)=e^{-2/3t}.$$

In this example the inverse transform is obtained directly. In the majority of applications, however, the inverse transform will not be so apparent and some work will be required before looking at standard transforms.

Example

$$x'(t)+4x(t)=8$$

$$x(0)=0$$

In this case the input to the equation is a constant and from the definition of the Laplace transform must be assumed to be zero for $t<0$. Thus the transform of the equation is given by

$$sX(s)+4X(s)=\frac{8}{s}.$$

The last term $8/s$ is obtained in the same way as the unit step function or can be considered to be 8. $1/s$, i.e. a scaled unit step.

Thus

$$X(s)=\frac{8}{s(s+4)}$$

If $X(s)$ is not one of the standard transforms it must therefore be expanded to obtain suitable factors which can be readily transformed back to the time domain. This is usually done using the method of partial fractions.

$$\frac{8}{s(s+4)} = \frac{2}{s} - \frac{2}{s+4}$$

Thus

$$x(t) = 2 - 2e^{-4t}.$$

A list of transforms is given in Table A.2.

A Summary of Useful Theorems

(i)   Linearity

Given two functions $f_1(t)$ and $f_2(t)$ then if

$$f_3(t) = Af_1(t) + Bf_2(t)$$

where A and B are constants

$$\mathscr{L}[f_3(t)] = \mathscr{L}[Af_1(t) + Bf_2(t)] = A\,\mathscr{L}[f_1(t)] + B\,\mathscr{L}[f_2(t)]$$
$$= AF_1(s) + BF_2(s). \tag{A.5}$$

This theorem has already been used by implication to obtain the transforms of the differential equations.

(ii)   Change of scale

$$\mathscr{L}[f(t/a)] = aF(as) \tag{A.6}$$

(iii)   The shift theorem

Consider a time function $f(t)$ specified at some time $(t-t_0)$, where t_0 is a constant, i.e. the time shift. The shifted function is now written as $f(t-t_0)$. The Laplace transform of this function is then given by

$$\mathscr{L}[f(t-t_0)] = \int_0^\infty f(t-t_0)e^{-st}\,dt.$$

Introducing a new variable x where

$$x = t - t_0$$

and since t_0 is constant, $dt = dx$, and also $t = x + t_0$ then

$$\mathscr{L}[f(t-t_0)] = \int_0^\infty f(x)e^{-s(x+t_0)}\,dx$$

or

$$\mathscr{L}[f(x)] = e^{-st_0}\int_0^\infty f(x)e^{-sx}\,dx.$$

Thus the Laplace transform of a time shifted function is given by

$$\mathscr{L}[f(t-t_0)] = e^{-st_0}F(s). \tag{A.7}$$

Example

Consider the time shifted step function shown in Fig. A.2

Now

$$\mathscr{L}[H(t)] = 1/s$$

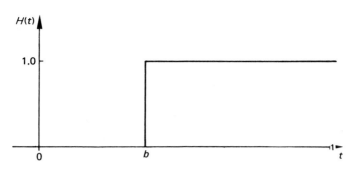

Fig. A.2

then
$$\mathcal{L}[H(t-b)] = \frac{1}{s}e^{-bs}$$

(iv)   Convolution
The concept of convolution has been covered in Chapter 5 and the same principles apply when using the Laplace transform, i.e.

$$\mathcal{L}\left[\int_0^t f_1(\tau)f_2(t-\tau)\mathrm{d}\tau\right] = F_1(s).F_2(s). \tag{A.8}$$

(v)   The initial value theorem
This theorem enables the value of the function $f(t)$, at time $t=0$, to be related to the Laplace transform of $f(t)$. To show this relationship consider the Laplace transform of the derivative of $f(t)$, $f'(t)$

$$\mathcal{L}[f'(t)] = sF(s) - f(0) = \int_0^\infty f'(t)e^{-st}\,\mathrm{d}t.$$

Now consider what happens when s tends to infinity, i.e. taking the limit as $s \to \infty$ on both sides of the equation yields

$$\lim_{s \to \infty} sF(s) - f(0) = \lim_{s \to \infty} \int_0^\infty f'(t)e^{-st}\,\mathrm{d}t.$$

Taking the limit inside the integration gives

$$\int_0^\infty f'(t)\lim_{s \to \infty} e^{-st}\,\mathrm{d}t = 0.$$

Thus
$$\lim_{s \to \infty} sF(s) - f(0) = 0$$

or
$$\lim_{s \to \infty} sF(s) = f(0). \tag{A.9}$$

This is the mathematical statement of the initial value theorem.

Example

Consider the Laplace transform model of the spring mass damper system discussed in Appendix C.

$$X(s) = \left[\frac{ms + C}{ms^2 + Cs + K}\right]x_i.$$

Applying the initial value theorem yields

$$\underset{s \to \infty}{\text{limit}} \, s \left[\frac{ms+C}{ms^2 + Cs + K} \right] x_i = x_i = x(0).$$

This is the initial value assumed in the derivation of the original model.

(vi)   Final value theorem

The final value theorem allows us to determine the value of $f(t)$, as t tends to infinity, from the Laplace transform of $f(t)$. Again we start with the transform of $f'(t)$ to obtain the required relationship but this time consider what happens when s tends to zero

$$\underset{s \to 0}{\text{limit}} \, sF(s) - f(0) = \underset{s \to 0}{\text{limit}} \int_0^\infty f'(t)e^{-st}dt.$$

Now take the limit $s \to 0$ inside the integration, thus

$$\int_0^\infty f'(t) \underset{s \to 0}{\text{limit}} \, e^{-st}dt = \int_0^\infty f'(t)dt$$

$$= [f(t)]_0^\infty$$

$$= \underset{t \to \infty}{\text{limit}} \, f(t) - f(0).$$

Therefore

$$\underset{s \to 0}{\text{limit}} \, sF(s) - f(0) = \underset{t \to \infty}{\text{limit}} \, f(t) - f(0)$$

i.e.

$$\underset{t \to \infty}{\text{limit}} \, f(t) = \underset{s \to 0}{\text{limit}} \, sF(s). \tag{A.10}$$

Example

Consider the Laplace transformed differential equation

$$(3s^2 + 2s + 1)X(s) = 3\,Y(s)$$

when the input $y(t)$ is a unit step function

$$\therefore X(s) = \frac{3}{3s^2 + 2s + 1} \cdot \frac{1}{s}.$$

To determine the final value of $x(t)$ as $t \to \infty$ we apply the final value theorem

$$\underset{t \to \infty}{\text{limit}} \, x(t) = \underset{s \to 0}{\text{limit}} \, X(s) = \underset{s \to 0}{s} \left[\frac{3}{3s^2 + 2s + 1} \cdot \frac{1}{s} \right]$$

$$= 3.$$

The output will tend towards a value of 3 units following a step input of one unit.

Appendix B

z-Transforms

The z-transform has been introduced in Chapters 5 and 8 and thus the purpose of this section is to formally derive some of the relationships used in the text. The z-transform can be thought of as the discrete equivalent of the Laplace transform. Indeed, if we replace the time variable (t) by the incremental time $(i\Delta)$ then the 'discrete' Laplace transform is written as:

$$F^*(s) = \sum_{i=0}^{\infty} f(i\Delta)e^{-i\Delta s} \qquad (B.1)$$

where $F^*(s)$ denotes the sampled (discrete) Laplace transform and the integral is replaced by a summation.

Now $f(i\Delta)$ is normally written as f_i and writing $z = e^{\Delta s}$ we have

$$F(z) = \sum_{i=0}^{\infty} f_i z^{-i}. \qquad (B.2)$$

This was of course derived in Chapter 5. The condition for convergence of this series is that

$$|z^{-1}| < 1 \quad \text{or} \quad |z| > 1.$$

Example

(i)   The unit step function

$$\begin{aligned} f_i &= 1 & i &= 0, 1, 2, \ldots, \infty \\ f_i &= 0 & i &= -1, -2, -3, \ldots, -\infty \end{aligned}$$

Thus
$$F(z) = \sum_{i=0}^{\infty} z^{-i}.$$

This version of the transform is not very useful and so the series is expressed in closed form. Initially this is obtained by multiplying both sides of the equation by z^{-1}

i.e.
$$z^{-1} F(z) = z^{-1} \sum_{i=0}^{\infty} z^{-i}$$

$$= z^{-1}(1 + z^{-1} + z^{-2} + \ldots + z^{-\infty})$$

$$= \sum_{i=1}^{\infty} z^{-i}.$$

Now subtract this equation from the original

$$F(z)(1 - z^{-1}) = \sum_{i=0}^{\infty} z^{-i} - \sum_{i=1}^{\infty} z^{-i} = 1$$

$$\therefore F(z) = \frac{1}{1 - z^{-1}}$$

$$\text{or } F(z) = \frac{z}{z - 1}.$$

(ii)   Now consider the discrete version of the time function

$$f(t) = e^{-at}$$

i.e.

$$f_i = e^{-ai\Delta}$$

Thus

$$F(z) = \sum_{i=0}^{\infty} e^{-ai\Delta} z^{-i}.$$

Note that this series will converge for values of z which satisfy

$$|e^{-ai\Delta} z^{-i}| < 1.$$

Again the series must be manipulated into closed form by multiplying both sides by $e^{-a\Delta} z^{-1}$ and subtracting the result from the same equation.

Thus

$$F(z) = \frac{1}{1 - e^{-a\Delta} z^{-1}}$$

$$= \frac{z}{z - e^{-a\Delta}}.$$

Notice in this case that the z-transform is dependent upon the sample interval which is normal when considering the z-transform of most signals and systems.

In addition to determining the z-transform of time functions we can also establish relationships between the Laplace and z-transforms of different functions

i.e.

$$\frac{1}{s + a} \leftrightarrow \frac{1}{1 - e^{-a\Delta} z^{-1}}$$

As with the Laplace transform we can construct a table—Table A.2.

The Solution of Difference Equations

Since the Laplace transform is one method of solving differential equations it is reasonable to expect the z-transform to be used in the solution of the discrete equivalent of the differential equation, i.e. the difference equation. However, we must first be clear about what is meant by the solution of a difference equation. By solution we mean the mathematical form of the equation such that the values of the dependent variable can be obtained without applying iteration,
i.e. if

$$y_{i+3} = -a_1 y_{i+1} - a_2 y_{i+2} + b_1 u_{i+1} + b_2 \mu_i \tag{B.3}$$

then for large values of the independent variable i a large number of iterations are required to achieve the appropriate value for y_i. As a consequence numerical errors are compounded and hence the final result may be significantly in error. To illustrate the method of solution consider the second order difference equation:

$$\alpha_2 y_{i+2} + \alpha_1 y_{i+1} + \alpha_0 y_i = \beta_1 \mu_{i+1} + \beta_0 \mu_i. \tag{B.4}$$

To determine the z-transform of terms such as y_{i+2} we must use the forward shift theorem (see section 5.5).

$$\mathscr{X}[y_{i+2}] = z^2 Y(z) - (z^2 y_0 + z y_1). \tag{B.5}$$

Similarly

$$\mathscr{X}[y_{i+1}] = z Y(z) - z y_0. \tag{B.6}$$

Note the similarity between these equations and the Laplace transform of derivatives.

Thus the z-transformed equation becomes:

$$\alpha_2(z^2 Y(z) - z^2 y_0 - z y_1) + \alpha_1(z Y(z) - z y_0) + \alpha_0 Y(z)$$
$$= \beta_1(z U(z) - z \mu_0) + \beta_0 U(z)$$

where

$$Y(z) = \mathscr{X}[y_i]$$
$$U(z) = \mathscr{X}[u_i]$$

and y_0, y_1 and μ_0 are the initial conditions.

Re-arranging the equation yields

$$(\alpha_2 z^2 + \alpha_1 z + \alpha_0) Y(z) = (\beta_1 z + \beta_0) U(z) + (\alpha_2 z^2 y_0 + \alpha_2 z y_1 + \alpha_1 z y_0)$$
$$- \beta_1 z \mu_0. \tag{B.7}$$

As in the case of the Laplace transform solution of differential equations the problem is simplified when all initial conditions are zero so we now assume y_0, y_1 and μ_0 to be zero to simplify the algebra. Re-arranging the equation once again:

$$Y(z) = \frac{\beta_1 z + \beta_0}{\alpha_2 z^2 + \alpha_1 z + \alpha_0} . U(z). \tag{B.8}$$

The final solution is now obtained by expanding this equation into partial fractions and using a table of transforms to obtain the time solution. Let us now consider a numerical example,

$$0.5 y_{i+1} - 0.45 y_i = 0.1 \mu_i$$

where the input sequence is a step of height 10. The z-transform of the difference equation is

$$0.5(z Y(z) - z y_0) - 0.45 Y(z) = 0.1 U(z).$$

Once again, assuming zero initial conditions

$$0.5 z Y(z) - 0.45 Y(z) = 0.1 U(z)$$

and

$$Y(z) = \frac{0.1}{0.5z - 0.45} . U(z).$$

Now

$$U(z) = \frac{10z}{z - 1} \quad \text{(z-transform of a step function)}$$

$$\therefore \ Y(z) = \frac{z}{(z - 1)(0.5z - 0.45)}$$

or

$$Y(z) = \frac{2z}{(z - 1)(z - 0.9)}.$$

Inspection of the table of transforms shows that this equation does not have an exact equivalent in the table, hence some algebraic manipulation is required,

i.e.
$$Y(z) = \frac{2}{1-0.9} \cdot \frac{z(1-0.9)}{(z-1)(z-0.9)}.$$

Now assuming
$$e^{-a\Delta} = 0.9$$

$$y_i = 20(1-0.9^i).$$

Hence y can now be calculated for any value of i without reference to any previous values, e.g.

$$y_4 = 20(1-0.9^4)$$

$$= 6.88.$$

Theorems of the z-transform

(i)   Addition and subtraction

$$\mathscr{Z}[f_1(t) \pm f_2(t)] = F_1(z) \pm F_2(z). \tag{B.9}$$

This can be shown to be true as follows

$$\mathscr{Z}[f_1(t) \pm f_2(t)] = \sum_{i=0}^{\infty} [f_1(i\Delta) \pm f_2(i\Delta)]z^{-i}$$

$$= \sum_{i=0}^{\infty} f_1(i\Delta)z^{-i} \pm \sum_{i=0}^{\infty} f_2(i\Delta)z^{-i}$$

$$= F_1(z) \pm F_2(z).$$

(ii)   Multiplication by a constant

$$\mathscr{Z}[af(t)] = \sum_{i=0}^{\infty} af(i\Delta)z^{-i}$$

$$= a \sum_{i=0}^{\infty} f(i\Delta)z^{-i}$$

$$= aF(z). \tag{B.10}$$

(iii)   The shift theorem
(a)   Backward shift
A function $f(t)$ shifted backwards in time is defined by the modified function

$$f(t-n\Delta)$$

where $n\Delta$ = the period of the backward shift.
Now by definition

$$\mathscr{Z}[f(t-n\Delta)] = \sum_{i=0}^{\infty} f(i\Delta - n\Delta)z^{-i}$$

$$= z^{-n} \sum_{i=0}^{\infty} f(i\Delta - n\Delta)z^{-(i-n)}.$$

Since we assume that $f(t)$ is zero when $t<0$ then the summation must only be valid from

$i = n$ to ∞

$$\therefore \mathscr{Z}\left[f(t-n\Delta)\right]=z^{-n}\sum_{i=n}^{\infty}f(i\Delta-n\Delta)z^{-(i-n)}$$

$$=z^{-n}F(z). \tag{B.11}$$

This means that the *z*-transform of a delayed sequence is obtained by deriving the *z*-transform of the original sequence and multiplying this by z^{-n} where n is the number of sample intervals required to represent the shift.

(b)   Forward Shift
A function which is shifted forward in time is given by $f(t+n\Delta)$

$$\mathscr{Z}\left[f(t+n\Delta)\right]=\sum_{i=0}^{\infty}f(i\Delta+n\Delta)z^{-i}$$

$$=z^{n}\sum_{i=0}^{\infty}f(i\Delta+n\Delta)z^{-(i+n)}$$

$$=z^{n}[F(z)-\sum_{i=0}^{n-1}f(i\Delta)z^{-i}]. \tag{B.12}$$

This is equivalent to the Laplace transform of a derivative. In this case the difference equation is obtained by applying forward differences to a continuous time function.

(iv)   The initial value theorem
As in the case of the Laplace transform we are interested in deriving the *z*-transform equivalent of the time function $f(t)$ when $t=0$, i.e. $f(0)$ is the initial value.
By definition

$$\mathscr{Z}\left[f(t)\right]=\sum_{i=0}^{\infty}f(i\Delta)z^{-i}$$

$$=f(0)+f(\Delta)z^{-1}+f(2\Delta)z^{-2}+f(3\Delta)z^{-3}+ \ldots$$

$$+f(\infty\Delta)z^{-\infty}.$$

Now when $z\to\infty$ $z^{-1}, z^{-2} \ldots z^{-\infty}$ become zero

$$\underset{z\to\infty}{\text{limit}}\, F(z)=f(0) \tag{B.13}$$

which is the initial value theorem for the *z*-transform.

(v)   The final value theorem
The final value theorem relates a time function $f(t)$ to a *z*-transform of the function as $t\to\infty$. To derive a relationship it is necessary, first of all to determine the *z*-transform of a function $f(n\Delta)$ and consider the special case of $n\to\infty$. To obtain $f(n\Delta)$ consider the two finite sequences

$$\sum_{i=0}^{\infty}f(i\Delta)z^{-i}=f(0)+f(\Delta)z^{-1}+ \ldots +f(n\Delta)z^{-n} \tag{B.14}$$

and

$$\sum_{i=0}^{n}f[(i-1)\Delta]z^{-i}=f(-1)+f(0)z^{-1}+f(\Delta)z^{-2}+ \ldots +f[(n-1)\Delta]z^{-n}.$$

Again we assume that $f(t)=0$ for $t<0$, hence the last equation becomes

$$\sum_{i=0}^{n} f[(i-1)\Delta]z^{-i} = f(0)z^{-1} + f(\Delta)z^{-2} + \ldots + f[(n-1)\Delta]z^{-n}$$

$$= z^{-1}\sum_{i=0}^{n-1} f(i\Delta)z^{-i}. \tag{B.15}$$

Subtracting equations (B.14) and (B.15) and noting that when $z \rightarrow 1$

$$\underset{z\rightarrow 1}{\text{limit}}\left[\sum_{i=0}^{n} f(i\Delta)z^{-i} - z^{-1}\sum_{i=0}^{n-1} f(i\Delta)z^{-i}\right] = \sum_{i=0}^{n} f(i\Delta) - \sum_{i=0}^{n-1} f(i\Delta).$$

$$= f(n\Delta).$$

We now require to apply the limit $n\rightarrow\infty$ in order to obtain the final value of the time function $f(t)$, (i.e. $f[\infty\Delta]$).
Hence

$$\underset{n\rightarrow\infty}{\text{limit}}\, f(n\Delta) = \underset{n\rightarrow\infty}{\text{limit}}\,\underset{z\rightarrow 1}{\text{limit}}\left[\sum_{i=0}^{n} f(i\Delta)z^{-i} - z^{-1}\sum_{i=0}^{n-1} f(i\Delta)z^{-i}\right].$$

Now we apply the limit $n\rightarrow\infty$ to the expression in the square brackets above, which enables the following equality to be established

$$\underset{n\rightarrow\infty}{\text{limit}}\sum_{i=0}^{n} f(i\Delta)z^{-i} = \underset{n\rightarrow\infty}{\text{limit}}\sum_{i=0}^{n-1} f(i\Delta)z^{-i} = F(z).$$

Thus we can write

$$\underset{n\rightarrow\infty}{\text{limit}}\, f(n\Delta) = \underset{z\rightarrow 1}{\text{limit}}\,[F(z) - z^{-1}F(z)]$$

or

$$\underset{n\rightarrow\infty}{\text{limit}}\, f(n\Delta) = \underset{z\rightarrow 1}{\text{limit}}\,(1 - z^{-1})F(z). \tag{B.16}$$

This equation is known as the final value theorem.

Appendix C

The Modelling of Physical Systems

In the context of this work modelling refers to the development of mathematical equations to represent the physical characteristics of a system or process. The mathematical model of a system can be developed in a number of ways, the most widely used methods being (a) the derivation of a model from physical laws and (b) the use of experimental data to derive a model. To illustrate this point consider the simple example shown in Fig. C.1. Given the source voltage V and the current i determine the resistance R. The physical model to represent this system is based on Ohm's law, which gives the relationship between v, i and R, i.e.

$$v = iR. \tag{C.1}$$

This is the model for the system from which R can be calculated. An alternative way of deriving a value for R in the absence of any other knowledge is to vary one of the parameters say v measure i and plot the results, see Fig. C.2. From these results we know

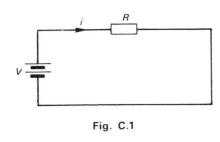

Fig. C.1

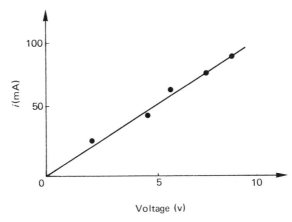

Fig. C.2

that

$$i = Kv \qquad \text{(C.2)}$$

and K is in fact $1/R$ because of the way in which the graph is drawn. Nevertheless, with this experimentally determined relationship we have derived a model for the system.

This example is very simple and relatively straightforward as none of the parameters vary with time but the essential stages in the development of a model remain the same for all systems.

In order to illustrate the modelling of dynamic systems and to give some background to some of the models used in the text the modelling of three different types of system will be considered.

Electrical systems

Example 1

As a development of the first electrical system, Fig. C.1, consider now the inclusion of a capacitor, Fig. C.3, and develop a mathematical model which will enable the change in voltage across the capacitor, v_0, to be determined following a change in input voltage v_i.

The derivation of a theoretical model is always based on some physical law or relationship. In this case we will use both Ohm's law and Kirchoff's circuit laws to develop a model.

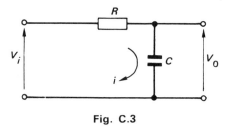

Fig. C.3

Kirchhoff's voltage law tells us that the sum of the voltage drops around a loop is equal to the applied voltage

$$v_i = \begin{pmatrix} \text{voltage drop across} \\ \text{resistor } R \end{pmatrix} + \begin{pmatrix} \text{voltage drop across} \\ \text{capacitor } C \end{pmatrix}$$

$$v_i = iR + v_0. \qquad \text{(C.3)}$$

But

$$v_0 = \frac{1}{C} \int_0^T i \, dt. \qquad \text{(C.4)}$$

This last equation being the voltage–current relationship for a capacitor. If we now take the Laplace transform of both equations, assuming all initial conditions are zero then

$$V_i(s) = Ri(s) + V_0(s)$$

and

$$V_0(s) = \frac{1}{C} \cdot \frac{I(s)^*}{s}.$$

* The Laplace transform of integration with zero initial conditions is $1/s$.

Since the input–output voltage relationship is required, $I(s)$ can be eliminated leaving the required voltage relationship,

i.e. $$V_i(s) = RCs\,V_0(s) + V_0(s)$$

or $$V_i(s) = (RCs + 1)V_0(s). \tag{C.5}$$

This equation is normally rearranged to give

$$V_0(s) = \frac{V_i(s)}{(RCs + 1)}$$

or $$\frac{V_0(s)}{V_i(s)} = \frac{1}{(RCs + 1)} = G(s) \tag{C.6}$$

which is known as the transfer function of the system. This equation can also be represented in block diagram form as shown in Fig. C.4.

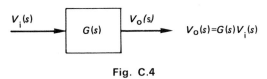

$$V_0(s) = G(s)V_i(s)$$

Fig. C.4

Example 2

More complex systems can be modelled in exactly the same way. Consider the electrical system in Fig. C.5. Again the relationship between V_0 and V_i is required.

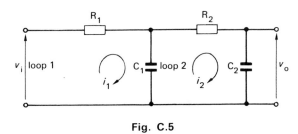

Fig. C.5

In this problem Kirchhoff's law must be applied to each of the current loops in turn.

$$v_i = i_1 R_1 + \frac{1}{C_1}\int_0^T (i_1 - i_2)\,dt \qquad \text{for loop 1} \tag{C.7}$$

$$\frac{1}{C_1}\int_0^T (i_1 - i_2)\,dt = i_2 R_2 + v_0 \qquad \text{for loop 2} \tag{C.8}$$

and $$v_0 = \frac{1}{C_2}\int_0^T i_2\,dt. \tag{C.9}$$

Taking the Laplace transform of these equations and substituting for I_2 we have

$$V_i(s) = R_1 I_1(s) + \frac{1}{C_1 s}(I_1(s) - C_2 s V_0(s)) \tag{C.10}$$

$$\frac{1}{C_1 s}(I_1(s) - C_2 s V_0(s)) = R_2 C_2 s V_0(s) + V_0(s). \tag{C.11}$$

Thus from equation (C.11) $I_1(s)$ can be determined from the other system variables

$$I_1(s) = s C_1 [R_2 C_2 s + C_2/C_1 + 1] V_0(s) \tag{C.12}$$

Substituting for $I_1(s)$ in equation (C.12) gives

$$V_i(s) = [R_1 R_2 C_1 C_2 s^2 + (R_1 C_2 + R_1 C_1 + R_2 C_2)s + 1] V_0(s)$$

or $\qquad \dfrac{V_0(s)}{V_i(s)} = \dfrac{1}{[R_1 R_2 C_1 C_2 s^2 + (R_1 C_2 + R_1 C_1 + R_2 C_2)s + 1]}. \tag{C.13}$

Both of these models can be transformed back to the time domain, thus giving the solution to the model for a specified input. See Appendix A for the solution of differential equations using the Laplace Transform.

Mechanical Systems

A typical mechanical system is shown in Fig. C6. The mass is considered to be in a state of equilibrium so that gravitational forces are balanced by the force in the spring due to the static deflection x_0,

i.e. $\qquad\qquad\qquad\qquad\qquad mg = K x_0.$

The gravitational force has now been eliminated from the problem so that Newton's second law can be applied to the mass, assuming a displacement x relative to the equilibrium position.

Newton's second law states that the inertia force is equal to the sum of the applied forces, i.e.

$$m\frac{d^2 x}{dt^2} = -Kx - C\frac{dx}{dt} \tag{C.14}$$

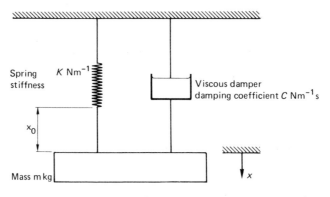

Fig. C.6

or
$$m\frac{d^2x}{dt^2} + C\frac{dx}{dt} + Kx = 0. \tag{C.15}$$

Now consider the transfer function for the system when the mass is displaced by a distance x_i and then released, i.e. we have now specified some initial conditions and so the Laplace transform must take this into account. Thus the transformed equation becomes

$$m(s^2X(s) - sx(0) - x'(0)) + C(sX(s) - x(0)) + KX(s) = 0.$$

Now $x(0) = x_i$ the initial displacement

and $x'(0)V_0 = 0$ the mass starts at rest (zero velocity) from the displaced position.

Therefore the transformed equation becomes

$$ms^2X(s) - msx_i + CsX(s) - Cx_i + KX(s) = 0.$$

Thus
$$X(s) = \frac{ms + C}{ms^2 + Cs + K} \cdot x_i. \tag{C.16}$$

Fluid Systems

Consider the interconnected tank system shown in Fig. C.7. The flow rate into tank 1 is Q_i and that out of tank 2 is Q_o. Clearly, as the inflow rate Q_i is increased the level in tank 1 will rise, the flow through valve 1 will increase and hence the level in tank 2 and outflow Q_o will increase. This can be expressed mathematically by using the continuity equation applied to each tank in turn, i.e.

Flow rate into a tank = flow rate out of a tank + rate of change of the volume of fluid in a tank

$$Q_i = Q_o + \frac{dV}{dt} \tag{C17}$$

where V is the volume of fluid.

But $V = AH$ where H is the level of the fluid in the tank. Therefore the equation can be rewritten as

$$Q_i = Q_o + A\frac{dH}{dt}. \tag{C.18}$$

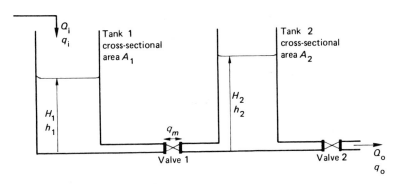

Fig. C.7

If we wish to model this system so that the level of liquid in each tank can be determined from a knowledge of the input then the flows Q_m and Q_o must be related to the level of liquid in each tank.

Now the flow through a valve can be approximated by

$$Q = a C_d \sqrt{2gH}$$

assuming the valve to be an orifice of cross-sectional area a, and discharge coefficient C_d

or

$$Q = K \sqrt{H}$$

where

$$K = a C_d \sqrt{2g}$$

which is a non-linear relationship. This relationship can be used directly but would lead to a non-linear model for the system. This aspect is outside the scope of the book, therefore some method of linearising the square root relationship must be devised.

The most common technique is to linearise the valve characteristic about some operating point, (i.e. some equilibrium point), and then consider small perturbations about this point. This can be done by calculating the slope or derivative of the function at the operating point, say H_o.

Now

$$\frac{dQ}{dH} \simeq = \frac{q}{h} = \frac{K}{2\sqrt{H}} = \frac{K}{2\sqrt{H_o}}\bigg|_{H=H_o} \tag{C.19}$$

$$\therefore \quad q = \frac{K}{2\sqrt{H_o}} . h$$

where q and h represent small deviations from the operating values—perturbation variables

or

$$q = K_p h \tag{C.20}$$

$$K_p = \frac{K}{2\sqrt{H_o}} .$$

Therefore the linearised valve characteristics become

$$q_m = K_1 (h_1 - h_2) \tag{C.21}$$

and

$$q_o = K_2 h_2 . \tag{C.22}$$

Thus the defining equations for each tank are:

Tank 1

$$q_i = K_1(h_1 - h_2) + A_1 \frac{dh_1}{dt} \tag{C.23}$$

Tank 2

$$K_1(h_1 - h_2) = K_2 h_2 + A_2 \frac{dh_2}{dt} . \tag{C.24}$$

Taking the Laplace transform of the equations yields

$$q_i = k_1(h_1(s) - h_2(s)) + A_1 s h_1(s) \tag{C.25}$$

$$k_1(h_1(s) - h_2(s)) = k_2 h_2(s) + A_2 s h_2(s) . \tag{C.26}$$

Now depending on the information required from the model there are three possible transfer functions. The first two transfer functions are obtained after some manipulation of the above equations which yields

$$\frac{h_1}{q_i}(s) = \frac{A_2 s + (K_1 + K_2)}{A_1 A_2 s^2 + (K_1 A_2 + K_1 A_1 + K_2 A_1)s + K_1 K_2} \tag{C.27}$$

$$\frac{h_2}{q_i}(s) = \frac{K_1}{A_1 A_2 s^2 + (K_1 A_2 + K_1 A_1 + K_2 A_1)s + K_1 K_2}. \tag{C.28}$$

By noting that $q_o = K_2 h_2$ the final transfer function can be written by inspection from equation (C.28)

$$\frac{q_o}{q_i}(s) = \frac{K_2 K_1}{A_1 A_2 s^2 + (K_1 A_2 + K_1 A_1 + K_2 A_1)s + K_1 K_2}. \tag{C.29}$$

Appendix D

A Fortran Subroutine to implement the FFT algorithm

```
C       SUBROUTINE TO COMPUTE THE FFT OF A COMPLEX TIME SERIES
C         AR,AI ARRAYS HOLDING REAL & IMAG. DATA
C         ( FOR REAL DATA AI SET TO 0 )
C         M = RADIX NUMBER
C         N = 2**M (MAX 1024)
C         ID = +1 TIME TO FREQUENCY TRANSFORMATION
C         ID = -1 FREQUENCY TO TIME TRANSFORMATION
C         NTIME = 1 COS & SIN TERMS COMPUTED
C         NTIME > 1 PREVIOUS VALUES OF COS & SIN USED
C
        SUBROUTINE FFT(AR,AI,M,N,ID,NTIME)
        DIMENSION AR(N),AI(N),ANGR(512),ANGI(512)
        NB2=N/2
        PI=4.0*ATAN(1.0)
C
C       TEST FOR DIRECTION OF TRANSFORMATION
C
        IF (ID.EQ.1) THEN
          DO 10 J=1,N
            AR(J)=AR(J)/FLOAT(N)
   10       AI(J)=AI(J)/FLOAT(N)
        END IF
        DI=FLOAT(ID)
C
C       COMPUTE COS & SIN TERMS IF NTIME=1
C
        IF (NTIME.EQ.1) THEN
          DO 20 I=1,NB2
            ALE=(I-1)*PI/FLOAT(NB2)
            ANGR(I)=COS(ALE)
   20       ANGI(I)=-SIN(ALE)*DI
        END IF
C
C       BIT REVERSE
C
        NM1=N-1
        J=1
        DO 30 L=1,NM1
          IF (L.LT.J) THEN
            T=AR(J)
            AR(J)=AR(L)
            AR(L)=T
            TX=AI(J)
            AI(J)=AI(L)
            AI(L)=TX
          END IF
          K=NB2
```

```
25        IF (K.LT.J) THEN
              J=J-K
              K=K/2
              GO TO 25
          END IF
30    J=J+K
C
C        COMPUTE FFT
C
      DO 50 M1=1,M
         ME=2**M1
         K=ME/2
         DO 50 J=1,K
            DO 40 L=J,N,ME
               LPK=L+K
               KI=(N/(2**M1))*(J-1)+1
               TR=AR(LPK)*ANGR(KI)-AI(LPK)*ANGI(KI)
               TI=AR(LPK)*ANGI(KI)+AI(LPK)*ANGR(KI)
               AR(LPK)=AR(L)-TR
               AI(LPK)=AI(L)-TI
               AR(L)=AR(L)+TR
               AI(L)=AI(L)+TI
40          CONTINUE
50    CONTINUE
      END
```

Basic Subroutine to Implement the FFT Algorithm

```
1000 REM BASIC CODE TO IMPLEMENT FFT
1010 REM AR = ARRAY HOLDING REAL DATA DIMENSION N
1020 REM AI = ARRAY HOLDING IMAG DATA DIMENSION N
1030 REM M  = RADIX NUMBER
1040 REM N  = 2^M
1050 REM ID = +1 TIME TO FREQUENCY TRANSFORMATION
1060 REM ID = -1 FREQUENCY TO TIME TRANSFORMATION
1070 IF ID>0 THEN FOR J=1 TO N
1080    AR(J)=AR(J)/N
1090    AI(J)=AI(J)/N
1100 NEXT
1105 REM
1110 REM BIT REVERSE
1115 REM
1120 NHLF=N/2
1130 NM1=N-1:J=1
1140 FOR L=1 TO NM1
1150    IF (L>=J) THEN 1200
1160    T=AR(J)
1170    AR(J)=AR(L):AR(L)=T
1180    TX=AI(J)
1190    AI(J)=AI(L):AI(L)=TX
1200    K=NHLF
1210    IF (K>=J) THEN 1250
1220    J=J-K
1230    K=K/2
1240    GOTO 1210
1250    J=J+K
1260 NEXT L
1270 REM
1280 REM FFT
1290 REM
```

```
1300 FOR M1=1 TO M
1310    UR=1.0:UI=0.0
1320    ME=2^M1
1330    K=ME/2
1340    CON=PI/K
1350    FOR J=1 TO K
1360      FOR L=J TO N STEP ME
1370        LPK=L+K
1380        TR=AR(LPK)*UR-AI(LPK)*UI
1390        TI=AR(LPK)*UI+AI(LPK)*UR
1400        AR(LPK)=AR(L)-TR
1410        AI(LPK)=AI(L)-TI
1420        AR(L)=AR(L)+TR
1430        AI(L)=AI(L)+TI
1440      NEXT L
1450    UR=COS(CON*J)
1460    UI=-SIN(CON*J)*ID
1470    NEXTJ
1480 NEXT M1
```

Solutions to Exercises

Exercises 1

1 $3\frac{1}{3}$V, 6.383V.

2 Mean value $= 1.743$
rms value $= 46.54$.

Exercises 2

3 $k = \frac{1}{16}$, $P(x \leqslant 6) = \frac{7}{8}$.

4 (a) is a PDD $\quad \bar{x} = (a+b)/2$
$\overline{x^2} = (b^2 + ab + a^2)/3$

    (b) is not a PDD

    (c) is a PDD $\quad \bar{x} = 0$
$\overline{x^2} = \frac{1}{2}$.

5 (a) and (b) are not PDDs
    (c) $\quad \bar{x} = \frac{2}{3}[0 + \frac{1}{3} + \frac{2}{9} + \frac{3}{27} +$
$\overline{x^2} = \frac{2}{3}[0 + \frac{1}{3} + \frac{4}{9} + \frac{9}{27} +$.

6 $\bar{x} = 0$, $\overline{x^2} = \frac{1}{2}$.

7 $\frac{1}{4}$, $\frac{2}{3}$.

8 (a) $\tilde{x} = -\frac{1}{4}$, $\quad \tilde{x}^2 = 5.25$, $\quad \sigma = 2.28$
    (b) 0.23.

Exercises 3

1 (a) $R_{xx}(\tau) = \left(\dfrac{1}{3} - \dfrac{\tau}{T} + \dfrac{\tau}{2T^2}\right)X^2 \quad 0 \leqslant \tau \leqslant 2T \qquad$ repeats period 2T

    (b) $R_{xx}(\tau) = (T - \tau)X^2 \qquad \left. \begin{array}{l} 0 \leqslant \tau \leqslant T \\ T < \tau \leqslant 3T \end{array} \right\}$ repeats period 3T.
$R_{xx}(\tau) = 0$

2 The first five values for the respective autocorrelations are
R_{xx} 2165, 1715, 1052, 617, 334
R_{yy} 1092, 1083, 1058, 1026, 988.

3 The first five values for cross-correlation are
R_{xy} $-172, -365, -493, -559, -587$
R_{yx} $-172, 35, 172, 254, 293$.

4 $R_{xy}(\tau) = XY\tau^2/2T$ $0 \leqslant \tau \leqslant T$
             $= XY[-3T/2 + 3\tau - \tau^2/T]$ $T < \tau \leqslant 2T$
             $= XY[3T - \tau]^2/2T$ $2T < \tau \leqslant 3T.$

7 $R_{uy}(\tau) = \dfrac{UY}{T}\left[\dfrac{(a+b)e^{-a\tau} - 2ae^{-b\tau}}{2a(b^2 - a^2)}\right]$ for $\tau \geqslant 0$

       $= \dfrac{UY}{T}\dfrac{e^{a\tau}}{2a(b+a)}$ for $\tau \leqslant 0.$

R_{uy} is periodic and this function is repeated for each period. The effect of the delay would be to delay the whole function by d for $d \leqslant T$.

8 $\bar{x} = 3,$ $\overline{x^2} = 16,$ $\sigma = 2.65,$ $f = 0.5$ Hz.

Exercises 4

3 $F(\omega) = \dfrac{-2j\omega}{\alpha^2 + \omega^2}.$

4 $F(\omega) = \dfrac{2\alpha}{\alpha^2 + \omega^2}.$

5 In Fig. 4.17 time function odd hence imaginary transform
In Fig. 4.18 time function even hence the real transform
$$F(\omega) = \frac{2(\alpha - j\omega)}{\alpha^2 + \omega^2} = \frac{2}{\alpha + j\omega}.$$

6

n	0	1	2	3	4	5	6	7
Real	0	0.250	0	0.250	0	0.250	0	0.250
Imaginary	0	0.604	0	0.105	0	−0.104	0	−0.604
Modulus	0	0.652	0	0.271	0,	0.270	0	0.652

7

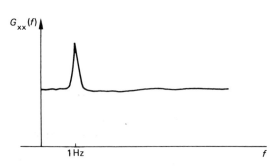

Fig. S4.7

8 Mean squared response is 20 units^2
$$R_{xx}(\tau) = \frac{2\sin(100\tau)}{(100\tau)} \text{ units}^2$$

Exercises 5

1 $y_{ss} = 0$

2 (i) $y(t) = \sqrt{50} \sin(2t + 45°) - 5e^{-2t}$
(ii) $y(t) = \sqrt{50} \sin(2t + 45°)$

3 For $t < -\tau$ $y = 0$
For $-\tau < t < \tau$ $y = \frac{1}{4} + \frac{1}{4}e^{-4(t+\tau)} - \frac{1}{2}e^{-2(t+\tau)}$
For $t > \tau$ $y = \frac{1}{2}e^{-2t}(e^{2\tau} - e^{-2\tau}) + \frac{1}{4}e^{-4t}(e^{-4\tau} - e^{4\tau})$

5 $y = 0$ $;t < -\tau$
$y = \frac{1}{4} + \frac{1}{4}e^{-4(t+\tau)} - \frac{1}{2}e^{-2(t+\tau)}$ $;-\tau < t \leqslant \tau$
$y = \frac{1}{2}e^{-2t}(e^{2\tau} - e^{-2\tau}) + \frac{1}{4}e^{-4t}(e^{-4\tau} - e^{4\tau})$ $;t > \tau$

6 $y_0 \simeq 0$, $y_1 \simeq 0.252$, $y_2 \simeq 0.574$, $y_3 \simeq 0.9$

7 (a) 0,   0,   0.253,   0.573,   0.896 approx.
(b) 0,   0,   0.248,   0.569,   0.892 approx.

9 (a) Unstable (b) Unstable (c) Stable (d) Stable (e) Unstable (f) Stable (g) Unstable
(h) Stable (i) Unstable (j) Unstable

11 $R_{uy}(\tau) = 0$ for $\tau < -0.1$
          $= 0.838 \, e^{-4\tau} - 2.047 \, e^{-2\tau} + 1.25$ for $-0.1 < \tau < 0.1$
          $= 1.0065 \, e^{-2\tau} - 1.027 \, e^{-4\tau}$ for $\tau > 0.1$

12 $R_{y_1 y_2}(\tau) = \displaystyle\int_{-\infty}^{\infty} \int_{-\infty}^{\infty} h_1(\mu) h_2(\lambda) R_{x_1 x_2}(\tau + \mu - \lambda) \, d\lambda \, d\mu$

14 (a) $R_{yy}(\tau) = \frac{5}{24}(e^{-2|\tau|} - \frac{1}{2}e^{-4|\tau|})$
(b) $R_{yy}(\tau) = 0.001315 \, e^{-20|\tau|} + 0.4208 \, e^{-2|\tau|} - 0.217 \, e^{-4|\tau|}$

Exercises 6

1 Mag. $= \sqrt{50}$ phase shift $= -45°$

4 (a) $G(z) = 4z/(z - 0.9048)$
(b) $G(z) = 0.0861z/(z - 0.9048)(z - 0.0861)$

5 (a) $G(z) = \dfrac{0.09516}{z - 0.9048}$

(b) $G(z) = \dfrac{0.0435(1 - z)}{(z - 0.9048)(z - 0.8187)}$

(Remember that Δ in the denominator of the TF of the ZOH cancels with that subsumed into the input signal and is thus ignored in frequency domain work).

6 $S_{uy}(\omega) = \dfrac{10}{(j\omega + 2)(j\omega + 4)}$; $S_{yy}(\omega) \dfrac{10}{(\omega^2 + 4)(\omega^2 + 16)}$.

7 $S_{uy}(\omega) = \dfrac{8000}{(j\omega + 2)(j\omega + 4)(\omega^2 + 400)}$

$S_{yy}(\omega) = \dfrac{8000}{(\omega^2 + 4)(\omega^2 + 16)(\omega^2 + 400)}$.

Exercises 7

1 The points to be on a circle (the unit circle) with the upper half being the complex conjugate of the lower half. The spectrum repeats around the unit circle.

2 Suggested cut-off frequency 300 Hz.
To give good signal reconstruction at 250 Hz choose the sample rate to be 1250 Hz
Range of analysis -625 Hz to $+625$ Hz.
For a resolution better than 1 Hz number of points 1250 (for FFT this means 2048 points).
If sequence padded with zeros, i.e. N_p points; N_z zeros then the resulting spectrum is scaled by the factor

$$\frac{N_p}{N_p + N_z}.$$

3

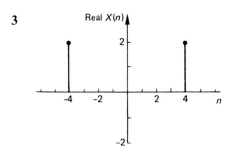
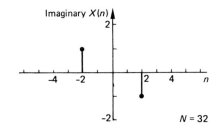

Fig. S7.3

4

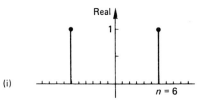
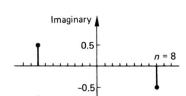

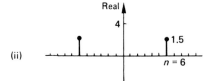
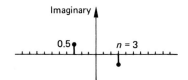

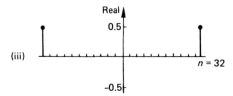
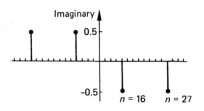

Fig. S7.4

5 $x(t) = 2 \sin 4\pi t + 4 \cos 8\pi t$.

6

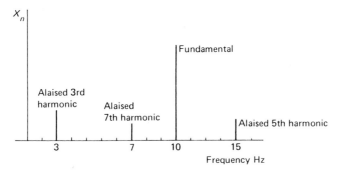

Fig. S7.6

7 Re-arranged sequence $\left\{ 0, 0, 1, -1, \dfrac{1}{\sqrt{2}}, -\dfrac{1}{\sqrt{2}}, \dfrac{1}{\sqrt{2}}, -\dfrac{1}{\sqrt{2}} \right\}$.

Exercises 8

2 (i) $x_n = 2 - (0.5)^n$ Steady state value $= 2$
 (ii) $x_n = \frac{2}{3} + \frac{1}{3}(-0.5)^n$ Steady state value $= \frac{2}{3}$.
Frequency response

$\omega\Delta$	0.1π	0.2π	0.3π	0.4π	0.5π	0.6π	0.7π	0.8π	0.9π
(i) $|G(j\omega)|$	3.345	2.267	1.510	1.063	0.800	0.641	0.544	0.468	0.450
(ii) $|G(j\omega)|$	0.454	0.486	0.544	0.641	0.800	1.063	1.510	2.267	3.340

 (i) Low pass filter.
 (ii) High pass filter.

3 Analogue cut-off frequency $= 100$ rad/s
 (i) Finite difference cut-off frequency $= 72$ rad/s
 (ii) Bi-linear cut-off frequency $= 92.5$ rad/s
 (iii) z-transform cut-off frequency $= 109$ rad/s
 4 passes required gives -23.2 dB@ 20 Hz.

4 $G(z) = \dfrac{0.0123(1 + z^{-1})}{(1 - 0.975 z^{-1})}$

 $y_n = 0.975 y_{n-1} + 0.0123(x_n + x_{n-1})$.

5 $G(z) = 0.0676 \dfrac{(1 + 2z^{-1} + z^{-2})}{1 - 1.14 z^{-1} + 0.41 z^{-2}}$

 $y_n = 1.14 y_{n-1} - 0.41 y_{n-2} + 0.0676(x_n + 2x_{n-1} + x_{n-2})$.

Table A1 $j\omega$ Transforms (Fourier)

$j\omega$-Transform	$f(t)$
1	$\delta(t)$
$\dfrac{1}{j\omega + a}$	$e^{-at}H(t)$
$\dfrac{1}{(j\omega + a)^2}$	$te^{-at}H(t)$
$\dfrac{-2}{\omega^2}$	$\lvert t \rvert$
$2\pi\delta(\omega)$	1
$\pi\delta(\omega) + \dfrac{1}{j\omega}$	$H(t)$
$\dfrac{\pi}{2}\left[\delta(\omega - \omega_0) + \delta(\omega + \omega_0) + \dfrac{j\omega}{\omega_0^2 - \omega^2}\right]$	$\cos\omega_0 t \cdot H(t)$
$\dfrac{\pi}{2j}\left[\delta(\omega - \omega_0) - \delta(\omega + \omega_0) + \dfrac{\omega_0}{\omega_0^2 - \omega^2}\right]$	$\sin\omega_0 t \cdot H(t)$
$\pi[\delta(\omega - \omega_0) + \delta(\omega + \omega_0)]$	$\cos\omega_0 t$
$j\pi[\delta(\omega + \omega_0) - \delta(\omega - \omega_0)]$	$\sin\omega_0 t$
$\dfrac{j\omega + a}{(j\omega + a)^2 + \omega_0^2}$	$e^{-at}\cos\omega_0 t \cdot H(t)$
$\dfrac{\omega_0}{(j\omega + a)^2 + \omega_0^2}$	$e^{-at}\sin\omega_0 t \cdot H(t)$
$2a \cdot \dfrac{\sin a\omega}{a\omega}$	$H(t + a) - H(t - a),\ a > 0$
$H(\omega + a) - H(\omega - a)$	$\dfrac{a}{\pi} \cdot \dfrac{\sin at}{at},\ a > 0$
$2a \cdot \dfrac{\sin^2 a\omega}{a^2\omega^2}$	$1 - \dfrac{\lvert t \rvert}{a},\ \lvert t \rvert < a$ $0,\ \lvert t \rvert > a$
$\dfrac{2a}{\omega^2 + a^2}$	$e^{-a\lvert t \rvert}$
$\dfrac{2\pi}{a}\displaystyle\sum_{n=-\infty}^{n=\infty}\delta\left(\omega - \dfrac{2\pi n}{a}\right)$	$\displaystyle\sum_{n=-\infty}^{n=\infty}\delta(t - na)$
$\dfrac{2}{j\omega}$	$\operatorname{sgn}(t)$

$H(t)$ is the unit step function

Table A2 s-Transforms (Laplace) & z-Transforms

s-Transform	$f(t)$ $[t=i\Delta$ for discrete]		z-Transform
1	$\delta(t)$	1 for $i=0$ 0 for $i\neq0$ discrete	1
$e^{-\tau s}[e^{-k\Delta s}$, discrete]	$\delta(t-\tau)$	1 for $i=k$ 0 for $i\neq k$ discrete	z^{-k}
$\dfrac{1}{s}$	$H(t)$	1 for $i\geqslant0$ 0 for $i<0$ discrete	$\dfrac{z}{z-1}$
$\dfrac{1}{s^2}$	t		$\dfrac{\Delta z}{(z-1)^2}$
$\dfrac{1}{s+a}$	e^{-at}		$\dfrac{z}{z-e^{-a\Delta}}$
$\dfrac{1}{(s+a)^2}$	te^{-at}		$\dfrac{\Delta z e^{-a\Delta}}{(z-e^{-a\Delta})^2}$
$\dfrac{1}{s(s+a)}$	$\dfrac{1}{a}(1-e^{-at})$		$\dfrac{z(1-e^{-a\Delta})}{a(z-1)(z-e^{-a\Delta})}$
$\dfrac{1}{s(s+a)^2}$	$\dfrac{1}{a^2}[1-e^{-at}(1-at)]$		$\dfrac{z(z(1-e^{-a\Delta}-a\Delta e^{-a\Delta})+e^{-2a\Delta}-e^{-a\Delta}+a\Delta e^{-a\Delta})}{a^2(z-1)(z-e^{-a\Delta})^2}$
$\dfrac{1}{(s+a)(s+b)}$	$\dfrac{1}{b-a}(e^{-at}-e^{-bt})$		$\dfrac{z(e^{-a\Delta}-e^{-b\Delta})}{(b-a)(z-e^{-a\Delta})(z-e^{-b\Delta})}$

Table A2 Continued

s-Transform	f(t) $[t=i\Delta$ for discrete]	z-Transform
$\dfrac{1}{s(s+a)(s+b)}$	$\dfrac{1}{ab}-\dfrac{e^{-at}}{a(b-a)}+\dfrac{e^{-bt}}{b(b-a)}$	$\dfrac{z[(a-b)(z-e^{-a\Delta})(z-e^{-b\Delta})+b(z-1)(z-e^{-a\Delta})-a(z-1)(z-e^{-a\Delta})]}{ab(a-b)(z-e^{-a\Delta})(z-e^{-b\Delta})(z-1)}$
$\dfrac{1}{s^2(s+a)}$	$\dfrac{1}{a^2}(at-1+e^{-at})$	$\dfrac{z[z(a\Delta-1+e^{-a\Delta})+(1-e^{-a\Delta}-a\Delta e^{-a\Delta})]}{a^2(z-1)^2(z-e^{-a\Delta})}$
$\dfrac{1}{s^2(s+a)^2}$	$\dfrac{1}{a^2}\left[t-\dfrac{2}{a}+\left(t+\dfrac{2}{a}\right)e^{-at}\right]$	$\dfrac{z\left\{(z-e^{-a\Delta})^2\left[\Delta-\dfrac{2}{a}(z-1)\right]+(z-1)^2\left(\Delta e^{-a\Delta}+\dfrac{2}{a}\right)(z-e^{-a\Delta})\right\}}{a^2(z-1)^2(z-e^{-a\Delta})^2}$
$\dfrac{1}{s^2(s+a)(s+b)}$	$\dfrac{1}{ab}\left(t-\dfrac{b+a}{ab}-\dfrac{a}{b(b-a)}e^{-at}+\dfrac{b}{a(b-a)}e^{-bt}\right)$	$\dfrac{z(K-L-M-N)}{a^2b^2(b-a)(z-1)^2(z-e^{-a\Delta})(z-e^{-b\Delta})}$
$\dfrac{s+a}{(s+a)^2+\omega^2}$	$e^{-at}\cos\omega t$	$\dfrac{z(z-e^{-a\Delta}\cos\omega\Delta)}{z^2-2e^{-a\Delta}z\cos\omega\Delta+e^{-2a\Delta}}$
$\dfrac{\omega}{(s+a)^2+\omega^2}$	$e^{-at}\sin\omega t$	$\dfrac{ze^{-a\Delta}\sin\omega\Delta}{z^2-2e^{-a\Delta}z\cos\omega\Delta+e^{-2a\Delta}}$
$\dfrac{1}{s[(s+a)^2+\omega^2]}$	$\dfrac{1}{a^2+\omega^2}\left[1-e^{-at}\left(\cos\omega t+\dfrac{a}{\omega}\sin\omega t\right)\right]$	$\dfrac{z(Az+B)}{(a^2+\omega^2)(z-1)(z^2-2e^{-a\Delta}z\cos\omega\Delta-e^{-2a\Delta})}$

$K=\Delta ab(b-a)(z-e^{-a\Delta})(z-e^{-b\Delta})$

$L=(b^2-a^2)(z-1)(z-e^{-a\Delta})(z-e^{-b\Delta})$

$M=a^2(z-1)^2(z-e^{-b\Delta})$

$N=b^2(z-1)^2(z-e^{-a\Delta})$

$A=1-e^{-a\Delta}\cos\omega\Delta-\dfrac{a}{\omega}e^{-a\Delta}\sin\omega\Delta$

$B=e^{-2a\Delta}+\dfrac{a}{\omega}e^{-a\Delta}\sin\omega\Delta-e^{-a\Delta}\cos\omega\Delta$

Index